Annual Conference of the American Forage and Grassland Council (AFGC 2021)

Savannah, Georgia, USA
3-5 January 2021

ISBN: 978-1-7138-2819-8

Printed from e-media with permission by:

Curran Associates, Inc.
57 Morehouse Lane
Red Hook, NY 12571

proceedings.com

Some format issues inherent in the e-media version may also appear in this print version.

Printed by Curran Associates, Inc. (2021)

American Forage and Grassland Council (AFGC)
PO Box 867
Berea, KY 40403
USA

Phone: 800.944.2342

www.afgc.org

Additional copies of this publication are available from:

Curran Associates, Inc.
57 Morehouse Lane
Red Hook, NY 12571 USA
Phone: 845-758-0400
Fax: 845-758-2633
Email: curran@proceedings.com
Web: www.proceedings.com

TABLE OF CONTENTS

2021 AFGC Virtual Annual Meeting Agenda

Workshop: 2021 Launch of ProClova Herbicide: Setting Expectations for Weed Control and Forage Tolerance in Pastures Interseeded with White Clover

Sponsored by Corteva Agrisciences

Coordiated, Moderated, and Presented by: Scott Flynn

ProClova, the first broadleaf weed herbicide labeled for preserving white clover in forage systems, will be available for purchase in the US in 2021. The goal of this workshop is to familiarize educators, distributors and end-users with the guidelines for proper use as well as to set the expectations on weed control and forage response after applications.

(Qualifies for 1.5 CEU)

Workshop: Animal Nutrition and Forages

Coordinated by: Matt Poore

Moderated by: Joy Hendrix

- Introduction – "Getting into the Weeds of Nutrition" by Dr. Matt Poore, NCSU
- Understanding Basic Cattle Nutrition by Dr. Deidre Harmon, NCSU
- Developing Cattle Nutrition Programs by Johnny Rogers, NCSU

(Qualifies for 1.5 CEU)

Workshop: Pasture Condition Scoring: Winter use and resource concerns

Held in collaboration with USDA NRCS

Coordinated by: Kevin Ogles, USDA NRCS

There are times when NRCS conservationists or producers need to conduct a pasture condition assessment during the winter. Can the Pasture Condition Scoring tool still be used with plants being dormant or dead during the non-growing season? Also, how do PCS indicator results guide the planner or producer to the resource concern on the pasture? And most of all, what can they advise the producer to do to solve that resource concern? This session will help answer these questions as well as provide an overview on the pasture condition scoring tool.

- Kevin Ogles, USDA NRCS, East National Technology Support Center
- Steve Woodruff, USDA NRCS, East National Technology Support Center
- Victor Shelton, USDA NRCS

(Qualifies for 1.5 CEU)

Volunteer Presentations

- Stem-to-Leaf Ratio: Recalculating Concepts of Forage Quality by Gonzalo Ferreira, Virginia Tech
- Mob and Rotational Grazing: Weighing Your Options by Eric Billman, USDA-ARS
- Forage Cell Wall Composition and Its Relationship with Undigested Neutral Detergent Fiber (uNDF) by Gonzalo Ferreira, Virginia Tech
- Mixed Perennial Grass-legume Swards Impact Deep Soil Characteristics by Victor Guerra, University of Florida

(Qualifies for 1.5 CEU)

Workshop: Forage Just Another Crop
Coordinated by: Johnny Rogers
- Forage Systems for Cow-Calf Production (Warm Season Grasses) by Dan Glenn, Deep Grass Graziers
- Forage Systems for Cow-Calf Production (Cool Season Grasses) by Ronnie Holman, Hardrock Beef Cattle
- Forage Systems for Stocker/Finisher Cattle by Paige Smart, Southeast AgriSeed

(Qualifies for 1.5 CEU)

Workshop: Soil Health
Coordinated and Moderated by: Alan Franzluebbers
A series of 10 minute presentations.
- Introduction to soil health – Alan Franzluebbers, USDA-ARS
- Seasonal changes of soil health in a perennial pasture overseeded with winter annuals by Kathleen Bridges (Louisiana State University), Lisa Fultz, Montgomery Alison, Kun Jun Han, William Pittman, and Bisoondat Macoon.
- Dung beetles affect nitrous oxide flux, ammonia volatilization, and nutrient cycling in grassland soils by Carlos Garcia (University of Florida), Jose Dubeux, Xavier Martini, Derrick Conover, Erick Santos, Bruno Grossi, Martin Ruiz-Moreno, Izabela Gomes, Luana Dantas, and Daciele Abreu
- Grazing cover crops: Impact of cover crop grazing on soil health in Alabama cropping systems by Russell Carrell (Auburn University), Audrey Gamble, Rishi Prasad, Kip Balkcom, Kim Mullenix, Yucheng Feng, and Leanne Dillard
- Traditional cropland cover crop or grazed double-crop forage impact on soil health by Jordan Cox-O'Neill (North Carolina State University), Deidre Harmon, Matt Poore, Alan Franzluebbers, Joan Eisemann, and Carrie Pickworth
- Do cattle cause compaction when grazing corn residue in the spring? by Morgan Grabau (University of Nebraska), Jim MacDonald, Zachary Carlson, and Mary Drewnoski
- Evaluating the effects of applied poultry litter and winter-fed hay on forage and soil biophysical properties by Lawrence Berry (University of Arkansas), Andrew Sharpley, and Ron Morrow
- Discussion period (20 minutes)

(Qualifies for 1.5 CEU)

Workshop: Calculated Parameters used in Forage Nutritive Analysis Reporting
Coordinated by: David McIntosh and Lisa Baxter
Moderator: Gary Bates
Panelist: Dr. Ralph Ward, Cumberland Valley Analytical Services; Dr. Dennis Hancock, USDA ARS; Kyle Taysom, DairyLand Laboratories; Neal Wininger, DairyLand Laboratories
David McIntosh, The University of Tennessee, Knoxville, NIRS Consortium

In forage nutritive analysis reporting there are a myriad of potential calculations available, however there is no guidance pointing towards the "best" option. This leads to large variability and confidence in these for example with Total Digestible Nutrients (TDN), Dry Matter Intake (DMI), and Relative Forage Quality (RFQ). There are numerous other calculated parameters that come to mind that this same issue is found when reporting is compared across labs and researchers. The Near-Infrared Spectroscopy Consortium (NIRSC) standardizes NIRS testing procedures, their application, and the accuracy of the constituents reported on a forage test. The National Forage Testing Association (NFTA) standardizes the accuracy of forage testing procedures. However, no organization regulates the calculated parameters used in the forage research and industry. The goal of this session is to have a moderated discussion on available calculations and determine what actions should be made in governing their use.
(Qualifies for 1.5 CEU)

Workshop: Making Hay
Sponsored by Massey Ferguson

Coordinated and Moderated by: Lisa Baxter

- Quality vs. Quantity by Leanne Dillard, Auburn University
- Best Use of Equipment for Quality Results by Lisa Baxter, University of Georgia
- New Technology Available to Hay Producers by Marcelo Wallau, University of Florida
- Southeastern Hay Contest Awards by SE Hay Contest Executive Committee

(Qualifies for 1.5 CEU)

Poster Presentations (OnDemand):

- Forage Quality of Mixed Systems in Organic Dairy Production by Ray Smith
- Estimating Soil Organic Carbon Sequestration of Pastures from Depth Distribution by Alan Franzluebbers
- A Forage-Based Educational Needs Assessment for Livestock and Forage Producers in Maryland by Amanda Grev
- Potassium Fertilization and its Impact on Yield, Nutritive Value, and Winter Survival of Alfalfa by Amy Greenberg
- Increasing Grassland Knowledge in WV with a Prescribed Grazing Professional Development Project by Ben Goff
- UK Master Grazer: Facilitating Change through On-Farm Demonstrations by Carrie Thrailkill
- Impact of Defoliation Height on Crown and Soil Temperature by Chris Teutsch and Jennifer Buckman
- The Effect of Irrigation Rate on the Yield and Species Composition of Two Perennial Grass/Legume Mixtures in Southern Oregon by Gordon Jones
- Establishing Southeastern Wild Rye: An Evaluation of Seeding Rate on Final Plant Population by Jesse Morrison
- The USDA Graze Pasture Budgeting Model for Kentucky- A New, Revisable Edition by Jimmy Henning
- Effect of Supplementation of Oilseeds to an Herbage-Based Diet High in Condensed Tannins on Enteric Methane Production and Ruminal Fermentation in Continuous Culture by Eric Billman and Kathy Soder
- Is There an Economic Advantage to Planting Diverse Summer Annual Forage Mixtures? by Kelly Mercier
- The History of the International Grassland Congress 1927-2020 by Ray Smith
- Mapping TNC Concentrations over Time in N-Fertilized and Unfertilized *Arachis Glabrata* Benth. Rhizomes by Kelly Thomas
- Novel Pest management Strategies to Control Smutgrass in Southern Forage Systems by Nicholas Shay
- Corn-Alfalfa Intercropping with Different Row Spacings by Samuel Bibby
- Wet Feet, Dry Feet, Hot Feet: Which Forages Might Tolerate all Three by Sharon Freeman
- Controlling Knotroot Foxtail in a Bermuda Hayfield by Terrell Davis
- Biomass Production and Forage Quality of Peanuts in the Southern Great Plains? by Travis Witt
- Fermentation and Aerobic Stability of Rain-Damaged Alfalfa/Grass Baled Silages Treated with a Propionic-Acid-Based Preservative by Wayne Coblentz
- White Lupin and Rye as Potential Winter Crop Forages for the Southeast by William Anderson
- Impact of Dry Matter at Ensiling on the pH and Nutritive Value of Baleage by Chris Teutsch and Alexander Teutsch
- Evaluation of Horse Pasture Score Card by Krista Lea
- Fall Grazing for Newly Weaned Lambs by Christine Gelley and Branden Campbell
- Application of Renote Sensing to Quantify Warm Season Grass Encroachment in Pasture by Jordyn Bush
- Meta-Analysis of the Effects of Chemical and Microbial Preservatives on Hay Spoilage by Marjorie Killerby

- Effect of Crabgrass and Summer Annual Legumes Mixes on Forage Yield and Nutritional Composition by Matias Aguerre
- Effects of Planting Population and Maturity at Harvest on Forage Yield and Nutritional Composition of Conventional and BMR Genotypes of Corn by O. M. Pena, G. Ferreira, and M. J. Aguerre
- Addressing the Status of Pasture Health through Producer Needs Assessment by Luke Harlow
- Effects of Defoliation and Perennial Peanut on Phosphorus and Micronutrients in Bahiagrass System by Adesuwa Erhunmwunse
- Impact of Fertilizer Type, Seeding Coating and Duration of Exposure on the Germination of Red Clover Seed by Chris Teutsch and Lindsay Sparks
- Impact of Fertilizer Type, Seeding Coating and Duration of Exposure on the Germination of Orchardgrass Seed by Chris Teutsch and Lindsay Sparks
- Sheep and Goat Preferences of Five Common Cover Crops by Robert Lane
- Yield Gains by Incorporating Rhizoma Peanut into Bahiagrass and its Impact on Soil Fertility Distribution by Jung-Chen Liu
- Water Quality for Georgia Livestock by Carole Knight
- Potential Use of Bermudagrass Turfgrass in Southern Agriculture Systems by Lisa Baxter
- Assessing the Impact of Agritultureal Machinery Traffic on Alfalfa Yields by Renee Anderson
- Comparison of Warm-Season Annual Grasses as Supplemental Grazing by Rocky Lemus
- Determing the Sphere of Influence of Spiny Amaranth (Amaranthus Spinosus) on Forage Yield Losses in Bermudagrass (Cynodon Dactylon) Pastures in Georgia:Weed Copmetition and Animal Avoidance by Tayler Danman

In-Person and OnDemand Volunteer Oral Presentations:

- Forage Mass and Structural Responses of Stockpiled Alfalfa-Bermudagrass Systems by Liliane Silva, Auburn University
- Pasture Management Curriculum for County Extension Agents by Caitlin Bainum, University of Florida
- On-Farm Demonstrations: Mistakes Made & Lessons Learned by Mark Mauldin, University of Florida
- Perceptions of Alfalfa Adoption in the Southeast US by L.S. Silva, M.K. Mullenix, J.J. Tucker, C.G. Prevatt
- Sericea Lespedeza--The Rodney Dangerfield of the Forage World by Reed Edwards
- An Update on the University of Tennessee Comprehensive Herbicide Stewardship Program by Neil Rhodes, The University of Tenneessee

*Continuing Education Units Total: **10.5** CEU's if attendee attended all the sessions during AFGC 2021 Virtual Conference

A FORAGE-BASED EDUCATIONAL NEEDS ASSESSMENT FOR LIVESTOCK AND FORAGE PRODUCERS IN MARYLAND

A.M. Grev[1] and S.B. Potts[2]

The objective of this project was to collect and formally document the educational needs, preferences, and challenges for livestock and forage producers in Maryland. A needs assessment survey was conducted online during April/May 2020 that was open to all livestock and forage producers and consisted of 36 questions designed to collect information on educational preferences, production and management practices, and forage-related challenges/successes. The survey was completed by 159 respondents spanning the dairy, beef, small ruminant, equine, and forage production industries. The majority of respondents were male (59%), 35 to 64 years old (70%), and had been involved in the industry for more than 11 years (61%). Rank-type questions were analyzed on a scale of 1 to 4, with 1 indicating no value/interest and 4 indicating a high value/interest. Private consultants (3.42) and regional sales representatives (3.09) received the highest rating as sources for forage-related information. The most preferred formats for receiving information included internet webpages (2.87), half-day sessions (2.85), and newsletters (2.83). Respondents showed the greatest interest in education on extending the grazing season (3.36), weed control (3.36), and improving soil organic matter (3.30). When asked to identify limitations to improving forage production, a lack of time/labor (2.44), high fertilizer costs (2.27), and dependence on hay (2.25) were listed as most limiting. Results of this assessment will be used to help Extension educators generate more effective forage-related educational programming to better meet the needs of the agricultural community in Maryland.

[1]Forage Extension Specialist and [2]Beef and Dairy Extension Specialist, University of Maryland Extension, Western Maryland Research and Education Center, Keedysville, MD.

ADDRESSING THE STATUS OF PASTURE HEALTH THROUGH PRODUCER NEEDS ASSESSMENT

L. Harlow, C.L. Mackowiak, A.R. Blount, M. Paret, and H.L. Liao[1]

Reports of pastures seemingly in decline but for undetermined reasons are on the rise these past several years. Pasture decline represents perennial forage stands that lose productivity and appear thin and weak without a simple, obvious cause. Pasture decline events have been reported throughout the lower Southeast Coastal Plain. Stand losses can vary in affected fields, causing losses in hay production, available grazing, seed and sod production. A needs assessment is being developed to help determine factors likely contributing to pasture decline. Initial steps include gaining a better understanding of the perception producers have about pasture decline and its possible relation to soil health factors. A survey was created to ascertain what management inputs are important to producers for the long-term health of their own perennial forage systems, if they had experienced pasture decline, and what management tools they either had used or are considering using to correct pasture decline and improve the overall health of their pastures and soil. The survey sent out via Qualitrics, has been distributed through social media marketing, as well as delivered to state cattlemen's association groups. Along with presenting preliminary results, we will address the citizen science approach to scientific inquiry to help fuel needs-driven research. Our long-term goal is to combine citizen-based, scientific inquiry with new extension educational programs to help identify and address complex problems.

[1]County Extension Director, University of Florida (UF)/IFAS Extension, Lake Butler, FL (harlow1231@ufl.edu); Soil Specialist, Soil and Water Sciences, UF, Quincy, FL (echo13@ufl.edu); Forage Breeder, Agronomy, UF, Quincy (paspalum@ufl.edu); Plant Pathologist, UF, Quincy (Paret@ufl.edu); Soil Microbiologist, Soil and Water Sciences, UF, Quincy, FL (sunny.liao@ufl.edu).

APPLICATION OF REMOTE SENSING TO QUANTIFY WARM SEASON GRASS ENCROACHMENT IN PASTURES

J. A. Bush[1] and C. D. Teutsch[2]

Remote sensing is becoming widely used in many areas of agriculture. Anecdotal evidence indicates that perennial cool season (C3) grass pastures may be experiencing encroachment by warm season (C4) grasses. The objective of this study was to determine if remote sensing technologies could be useful for detecting and monitoring this encroachment. These grasses have alternate periods of seasonal dormancy, which are accompanied by color change. During cooler periods C3 grasses will be greener while C4 grasses will be dormant and brown. This study analyzed the use of digital camera imagery and drone imagery to measure strips of brown (dead) and green (alive) grass in a pasture. A randomized complete block design with three replications was used. Treatments were 25, 50, and 75% of total area sprayed with non-selective herbicide. In April of 2020, strips were sprayed out of paddocks that had been planted to wheat. Ground measurements were used to confirm the actual treatment percentages. Thirty digital images were collected in a random pattern in each paddock. Green canopy cover for these images were determined using Canopeo software (OSU, Stillwater, OK). Regression equations for the digital camera imagery on two collection dates were $y = 0.0488 + 0.7497x$ and $y = 0.1749 + 0.6165x$, both having $p < 0.001$, and correlation coefficients of $r^2 = 0.9877$ and 0.9559, respectively. This initial study demonstrates that digital imagery may be a time-efficient, cost-effective method for monitoring the long-term shifts in the ratios of cool- to warm-season grasses in pastures in the transition zone of the United States.

[1]Graduate Research Assistant, University of Kentucky, Lexington, KY (jordynbush@uky.edu);
[2]Extension Associate Professor, University of Kentucky Research and Education Center, Princeton, KY (chris.teutsch@uky.edu).

ASSESSING THE IMPACT OF AGRICULTURAL MACHINERY TRAFFIC ON ALFALFA YIELDS

R.A. Anderson, N. Gaur, and J.D. Hale[1]

Alfalfa is a perennial forage crop widely grown within the United States. Vehicular traffic on alfalfa fields during harvest can damage alfalfa crowns as well as compact the soil depending on the wheel track of the vehicle. This damage has been shown to negatively impact its yield and nutritive value in subsequent harvests. This study simulated the effect of vehicular traffic on two varieties of alfalfa ('AmeriStand 435TQ RR' and 'Alfagraze 600RR') and soil during harvest and evaluated its impact on alfalfa yield. The experiment was conducted in a greenhouse setting using a machine that applies 207 kPa of pressure with 5 timings of application to simulate machinery traffic within both silage and hay harvest systems. Results showed that there was a decrease in yield as the season went on, but more analysis is needed to determine if compaction type or timing influenced the decrease. We expect that the results from this study will be valuable to allow for modeling of changes in yield and nutritive value under different machinery traffic management scenarios.

[1]Graduate Research Assistant, University of Georgia, Athens, GA (ra63515@uga.edu); Assistant Professor, University of Georgia, Athens, GA; Research Technician, University of Georgia, Watkinsville, GA.

BUILDING AND EXECUTING A HIGH-QUALITY FORAGE CHAIN

Terry Chandler[1]

Stillwater Farm is a cow-calf and stocker operation in Danielsville, GA on approximately 130 acres. Management-Intensive-Grazing is utilized on more than 28 permanent paddocks. The forage base is predominantly Russell and Tift-44 hybrid bermudagrasses with a crabgrass component where these stands have thinned. Bermudagrass fields are overseeded sequentially each fall with annual ryegrass, small grains, and annual clovers either in monoculture or mixtures to provide fall and winter grazing. Forage brassicas are utilized each fall for quick, high-quality grazing that supplements stockpiled bermudagrass. Brassicas are often followed by summer annual forage crops. These annual rotations have recently been utilized to facilitate conversion of some paddocks to novel-endophyte tall fescue varieties. The cool season perennial component of the operation has also been expanded with recent establishment of a hardwood silvopasture. Red and white clover have been introduced primarily though winter frost-seeding.

[1] Operator, Stillwater Farm, Danielsville, GA (tdchandler@windstream.net).

CALCULATED PARAMETERS USED IN FORAGE NUTRITIVE ANALYSIS REPORTING

L. Baxter, G. Bates, D. Combs, D. Hancock, R. Lemus, D. McIntosh, K. Taysom, D. Undersander, and R. Ward[1]

In forage nutritive analysis reporting there are a myriad of potential calculations available, however there is no guidance pointing towards the "best" option. This leads to large variability and confidence in these; for example: Total Digestible Nutrients (TDN), Dry Matter Intake (DMI), and Relative Forage Quality (RFQ), and Relative Feed Value (RFV). There are numerous other calculated parameters that come to mind where this same issue is found when reporting is compared across labs and researchers. Variability and confusion can be found in research articles, industry reports, commercial lab reports, popular press articles, and used more predominantly for pricing hay in domestic and international markets. This is a convoluted, yet pressing problem in the forage industry that needs to be addressed. However, no organization monitors and determines if the calculated parameters used in the research and industry are used appropriately. The goal of this session is to have a moderated panel discussion on only a small portion of available calculations to determine what actions should be made in standardizing their appropriate use.

[1]Assistant Professor and Extension Forage Agronomist, University of Georgia, Tifton, GA (baxterl@uga.edu); Professor and Director, University of Tennessee Beef and Forage Center, Knoxville, TN (gbates@utk.edu); Professor, University of Wisconsin, Madison, WI (dkcombs@wisc.edu); Center Director, USDA-ARS, U.S. Dairy Forage Research Center, Madison, WI (Dennis.Hancock@usda.gov); Professor and Extension Forage Specialist, Mississippi State University, Mississippi State, MS (rlemus@pss.msstate.edu); Coordinator, University of Tennessee Beef and Forage Center, Knoxville, TN (dmcintos@utk.edu); Business Development Manager, Dairyland Laboratories Inc, Arcadia, WI (ktaysom@dairylandlabs.com); Forage Professor Emeritus, University of Wisconsin, Madison, WI (djunders@wisc.edu); President, Cumberland Valley Analytical Services, Waynesboro, PA (rward@foragelab.com).

COMPARISON OF WARM-SEASON ANNUAL GRASSES AS SUPPLEMENTAL GRAZING

T. Peavey, R. Lemus, and J.D. Rivera, and J.A. White[1]

Profitability of beef cattle production in the southern USA relies on highly nutritive pastures. Traditional forage species in this region, such as bahiagrass (*Paspalum notatum*) and bermudagrass (*Cynodon dactylon*) tend to have limited nutritive value. Warm-season annual grasses (WSAG) such as crabgrass (*Digitaria spp.*) and pearl millet (*Pennisetum typhoides*) can be used as a supplemental forage to improve animal performance during summer months. The objective of this study was to determine the impact of WSAG's compared to bahiagrass on forage DM yield, nutritive value, and performance of growing beef cattle. The study contained five treatments replicated three times. Treatments included pearl millet ('Prime 380'), three crabgrass cultivars [Mojo Brand ('Red River' + "Impact'), 'Red River,' and 'Quick-N-Big Spreader'), and bahiagrass. Pearl millet was planted at 25 lb PLS/acre and crabgrass at 10 lb. Treatments were fertilized with 100 lb N/acre/yr. Pre-grazing forage samples were collected to determine forage biomass and nutritive value. Grazing began when pearl millet reached 18 to 14 inches and crabgrass was 12 to 15 inches. The average stocking rate was 1500 lb live weight/acre. Total biomass production was similar among all treatments. Bahiagrass had the lowest nutritive value compared to WSAG during the same grazing duration. Overall, WSAG produced 80% greater ADG compared to bahiagrass. Red River crabgrass produced greater gain per acre (112 lb/acre) among treatments.

[1]Undergraduate Apprentice Student, Mississippi State University, MS (mtp161@msstate.edu); Professor and Extension Forage Specialist, Mississippi State University, MS (Rocky.Lemus@msstate.edu); Associate Research/Extension Professor, Mississippi State University, MS (Daniel.Rivera@msstate.edu); Forage Variety Testing Manager, Mississippi State University, MS (JWhite@pss.msstate.edu).

CONTROLLING KNOTROOT FOXTAIL IN A BERMUDAGRASS HAYFIELD

T.A. Davis, B. Griffin[1]

Knotroot foxtail is an invasive perennial warm season weed that is quickly overtaking Arkansas pastures and hayfields. The main concerns with Knotroot foxtail include the prolific seed head, which produces many seeds and often cause abscesses in the gums of horses and other grazing livestock, and the extensive root system, which can make control difficult once the weed is established in pastures in hayfields. Many recommendations exist for annual foxtails, but few are found for perennial foxtails such as Knotroot. The objective of this demonstration was to compare the effectiveness of several non-selective herbicides that are readily available to area producers to control Knotroot foxtail. Johnson County Staff Chair Blair Griffin created a demonstration plan and established a demo on July 21, 2020 in a bermudagrass hayfield in Pike Co. Arkansas with cooperator Cody Self. Five plots compared the effectiveness of Paramount at 32 oz/acre (plot 1), Roundup up 12 oz/acre plus Patriot at 0.5 oz/acre (plot 2), Roundup at 12 oz/acre (plot 3), Roundup at 8 oz/acre plus Pastora at 1 oz/acre (plot 4), and Pastora at 1.5 oz/acre (plot 5). Plots were rated on September 8, 2020 for control effectiveness plot 1- 60%, plot 2- 95%, plot 3-95%, plot 4-90%, plot 5-90%. In previous studies from 2012-2019, Roundup at 16 oz/acre (90%) and Round at 16 oz/acre plus Patriot at 0.5 oz/acre (85%) demonstrated the most effective control.

[1]Staff Chair/CEA-Ag, University of Arkansas Cooperative Extension Service, Pike County, Murfreesboro, AR (tadavis@uaex.edu); Staff Chair/CEA-Ag, University of Arkansas Cooperative Extension Service, Johnson County, Clarksville, AR (bgriffin@uaex.edu) .

CORN-ALFALFA INTERCROPPING WITH DIFFERENT ROW SPACINGS

S., Bibby, M.T. Berti, A.F. Wick, D.P. Horvath, J.V. Anderson,
A. Wittenberg, A. Greenberg, and K. Mozea[1]

Alfalfa is a staple crop grown by many dairy and beef farmers as well as some crop farmers in the northern Great Plains. The seeding year for alfalfa is almost always the least productive and often only provides one to two cuttings in the northern Great Plains. To determine if seeding alfalfa with corn could increase alfalfa yield and nutritive value during the second year, as well as providing a corn crop during the first year, experimental studies were conducted in Prosper and Hickson, ND. The experimental design was a randomized complete block with four replicates at each location. Treatments included alfalfa alone, corn alone at 30- and 60-inch row spacings and corn intercropped with alfalfa at 30- and 60-inch row spacings. Alfalfa and corn were seeded on 21 May 2020. The overall scale of success within these treatments is measured in corn yield in the first year and alfalfa yield in year two. Corn grain and alfalfa biomass collected in the first year of the study was analyzed for composition with near infrared spectrometry. Data collected on each experimental unit also included: water stress, weed suppression, and soil health. The data sets collected over this two-year study will allow for more specific recommendations regarding alfalfa-corn intercropping. Optimizing this specific cropping system for growers in the northern Great Plains could increase profitability as well as forage nutritive value and crop efficiency.

[1]Graduate Research Assistant, MS student (samuel.bibby@ndus.edu); Forage and Biomass Production, Professor (marisol.berti@ndsu.edu); Associate Professor, Ext. Soil Health Specialist (abbey.wick@ndsu.edu); Research Plant Physiologist (david.horvath@usda.gov); Research Chemist (james.v.anderson@usda.gov); Research Specialist (alex.1.wittenberg @ndsu.edu); MS student (amy.greenberg@ndsu.edu); MS student (kenneth.mozea@ndsu.edu) North Dakota State University, Fargo, ND.

DETERMINING THE SPHERE OF INFLUENCE OF SPINY AMARANTH (*AMARANTHUS SPINOSUS*) ON FORAGE YIELD LOSSES IN BERMUDAGRASS (*CYNODON DACTYLON*) PASTURES IN GEORGIA: WEED COMPETITION AND ANIMAL AVOIDANCE

T.D. Denman[1], N.T. Basinger[2], and J.D. Hale[3]

Some weeds can have a detrimental effect on forages through direct competition for nutrients and light. Weeds can also influence grazing of forages as livestock deliberately avoid forages surrounding some weeds, leaving valuable forage untouched. The purpose of this study is to: 1.) Determine the competition of spiny amaranth on bermudagrass yield and 2.) Evaluate bermudagrass yield loss due to livestock avoidance of forage surrounding spiny amaranth. Four spiny amaranth plant densities were evaluated (0, 1, 2, and 4 plants per ft^2) at two harvest timings (early season and late season) in two traditional bermudagrass practices (hayfield and pasture). Spiny amaranth densities were established in the center of each plot and yield measurements were obtained at four distances (1, 2, 3, and 4 ft) away from the center of each plot at the designated harvest timing. Results indicate spiny amaranth yield was greater in the pasture study than in the hayfield study. This indicates that at these plant densities on a well-maintained hayfield, spiny amaranth may have limited impact on bermudagrass growth. In a less manicured common bermudagrass pasture, spiny amaranth could pose more of a competitive advantage. There were no discernable trends in the early season harvest between the bermudagrass yield and the distance from the spiny amaranth in either the hayfield or the pasture study; however, some trends for late season indicate that there may be an impact on grazing avoidance.

[1]Graduate Research Assistant, University of Georgia, Athens, GA; [2]Assistant Professor, University of Georgia, Athens, GA; [3]Research Technician, University of Georgia, Athens, GA.

EFFECT OF CRABGRASS AND SUMMER ANNUAL LEGUMES MIXES ON FORAGE YIELD AND NUTRITIONAL COMPOSITION

M. J. Aguerre and O. Peña[1]

Warm season annuals can complement perennial cool-season forage systems and extend production during summer months. Crabgrass is a high-quality summer forage that can support good animal performance, but it might not consistently fulfil the high nutrient requirements of stocker and dairy cattle. Summer annual legumes have the potential to provide high forage quality and additionally fix atmospheric N. We hypothesize that mixing annual legumes with crabgrass will increase forage yield and nutrient digestibility. Crabgrass (CG), CG + N (50 lbs./acre of N after each cut); cowpea (CWP) and Lablab (LL) were planted in monoculture (5, 50, and 20 lb/acre, respectively) or in mixtures (1/2 + 1/2 or 1/3 + 1/3 + 1/3 of monoculture seed density) resulting in 7 treatments replicated four times. Whole plots were harvested when crabgrass reached heading stage and samples were collected from each plot for nutrient analyses and in vitro digestibility. Highest yield per acre (as-is) obtained after 3-cuts was observed for CG (17,782 lb), lowest for LL (8,102 lb), and intermediate for all other treatments (14,778 lb). In addition, the majority of harvested biomass was obtained in the first cut (63%), with a smaller proportion in the second (24.5%) and third (13.8%) cuts, which reflected little regrowth of the summer legumes. Excessive drought was experienced during the summer and may have affected yield responses.

[1]Assistant Professor, Animal and Veterinary Science Department, Clemson University, Clemson, SC (maguerr@clemson.edu); Graduate student, Animal and Veterinary Science Department, Clemson University, Clemson, SC (openape@g.clemson.edu).

EFFECT OF PLANTING POPULATION AND MATURITY AT HARVEST ON FORAGE YIELD AND NUTRITIONAL COMPOSITION OF CONVENTIONAL AND BMR GENOTYPES OF CORN

O. M. Peña, G. Ferreira, and M. J. Aguerre[1]

Previous studies have shown that increasing planting population may increase corn silage yield without impacting nutrient composition and digestibility. However, the effect may depend on crop maturity at harvesting time. Thus, the objective of this study was to evaluate the effects of corn planting population, using conventional (CONV) and brown midrib (BMR) genotypes, and maturity stage at harvest on forage dry matter (DM) yield and silage quality and digestibility. The study was conducted in 2 fields (Location 1 and 2), where CONV and BMR corn genotypes were planted at a theoretical planting population of 24,000, 32,000, and 40,000 seeds/acre (24K, 32K, and 40K, respectively). Corn varieties were established on May 4 (Location 1) and June 9 (Location 2), whole plots harvested at either early-dent or 3/4 milk-line stage, and samples were collected from each plot for nutrient analyses. Average DM yield in Location 1 was 57% lower than in Location 2, and there was no yield difference between corn genotypes. Planting corn at 24K resulted in the lowest DM yield (10,406 lb/acre), whereas DM yield was similar for 32K and 40K (11,872 lb/acre). Yield of early harvest corn was lower compared with corn harvest at a more advanced maturity (10,605 vs. 12,163 lb/acre, respectively), but no interaction was observed between maturity and planting density. Results of this study suggest that regardless of planting population, maturity at harvest had minimal impact on corn DM yield.

[1]Graduate student, Animal and Veterinary Science Department, Clemson University, Clemson, SC (openape@g.clemson.edu); Dairy Management Extension Specialist, Department of Dairy Science Research, Virginia Tech, Blacksburg, VA (gonf@vt.edu); Ruminant Nutritionist, Animal and Veterinary Sciences Department, Clemson University, Clemson, SC (maguerr@clemson.edu).

EFFECT OF SUPPLEMENTATION OF OILSEEDS TO AN HERBAGE-BASED DIET HIGH IN CONDENSED TANNINS ON ENTERIC METHANE PRODUCTION AND RUMINAL FERMENTATION IN CONTINUOUS CULTURE

Eric D. Billman[1], Ana I. Roca-Fernández[2], S. Leanne Dillard[3], and Kathy J. Soder[4]

Compounds in plants, such as condensed tannins (CT), may reduce methane (CH_4) production in ruminants, but have negative impacts on ruminal fermentation. Supplementation with oilseeds may maintain ruminal fermentation, while further reducing methane. A 4-unit continuous culture fermentor system was used to assess nutrient digestibility, volatile fatty acid (VFA) production, bacterial protein synthesis and CH_4 production of various sources of oilseeds to improve nutrient digestibility of high-CT diets while further reducing enteric CH_4 production. The experimental design was a 4 × 4 Latin square, with 4 diets: a basal diet of 45% orchardgrass and 45% CT-containing sericea lespedeza plus 10% oilseed of either canola (CAN), soybean (SOY), sunflower (SUN), or equal proportions of all three oilseeds (MIX). Samples for dry matter, organic matter, crude protein, acid detergent fiber, and neutral detergent fiber were collected to determine nutrient digestibilities. Bacterial protein synthesis and VFA analyses were evaluated. Gas samples for CH_4 analysis were recorded by a photoacoustic field gas monitor. The SOY diet had greater ($P < 0.001$) CH_4 production than any of the other oilseeds. Nutrient digestibilities and bacterial N metabolism or flows were not affected ($P > 0.05$) by oilseed source. The addition of canola and sunflower seed meal to forages high in CT may provide excellent reductions in daily CH_4 emissions, while maintaining desirable feed digestibility in cattle.

[1]Research Agronomist; USDA-ARS, Florence, SC (eric.billman@usda.gov); [2]Postdoctoral Researcher, Universidad de Santiago de Compostela, Lugo, Spain (anai.roca@usc.es); [3]Assistant Professor/Extension Forage Systems Specialist, Auburn University (dillasa@auburn.edu); [4]Research Animal Scientist, USDA-ARS, University Park, PA (kathy.soder@usda.gov).

EFFECTS OF DEFOLIATION AND PERENNIAL PEANUT ON PHOSPHORUS AND MICRONUTRIENTS IN BAHIAGRASS SYSTEM.

A.S. Erhunmwunse, L.D. Queiroz, C.L. Mackowiak, J.C.B. Dubeux Jr., A.R.S. Blount, and H.L. Liao[1]

Rhizoma peanut (RP) (*Arachis glabrata* Benth.), a warm-season perennial legume incorporated into bahiagrass (*Paspalum notatum* Flüggé) pastures has been shown to increase forage productivity. However, there is limited information on the effect this mixture has on phosphorus (P) and micronutrient cycling. A field study was conducted at Marianna, FL to evaluate P and micronutrient distributions of bahiagrass grown as monoculture and as mixtures with 'Ecoturf' (EPS-BG) or 'Florigraze' (FPS-BG) RP in defoliated and non-defoliated systems. Exclusion cages of 1 x 1 m were installed in each plot to exclude grazing, while cattle grazed the remaining area, according to producer practices. All treatments were mowed monthly. Shoot, root+rhizome, and soil P, Fe, Zn, Cu, and Mn concentrations were measured. Shoot P were 47% greater in bahiagrass grown as an RP mixture (1.5 g kg^{-1}) than grown in monoculture (0.8 g kg^{-1}). Micronutrients were unaffected by forage species except shoot Fe which was lower in bahiagrass grown as RP mixture than as monoculture. Regardless of the plant mixture, bahiagrass and RP shoot Zn concentrations and root+rhizome Cu and Mn concentrations were significantly reduced under defoliation. However, forage species had no measurable effect on Mehlich-3 extractable soil P or micronutrients. We will further address the effect of forage composition and defoliation on soil P forms (fractions) in 2021.

[1]Graduate Student, Soil and Water Sciences, University of Florida (UF), Quincy, FL (erhunmwunsea@ufl.edu); Graduate Student, Agronomy, UF, Marianna, FL (ldantasqueiroz@ufl.edu); Soil Specialist, Soil and Water Sciences, UF, Quincy, FL (echo13@ufl.edu); Forage Management Specialist, Agronomy, UF, Marianna, FL (dubeux@ufl.edu); Forage Breeder, Agronomy, UF, Quincy (paspalum@ufl.edu); Soil Microbiologist, Soil and Water Sciences, UF, Quincy, FL (sunny.liao@ufl.edu).

ESTABLISHING SOUTHEASTERN WILD RYE: AN EVALUATION OF SEEDING RATE ON FINAL PLANT POPULATION

J.I. Morrison[1]

Southeastern wildrye is a native, cool-season perennial grass that has utility in monoculture and grass-legume mixtures planted for forage production in the southeast United States. Southeastern wildrye is susceptible to extended periods of heat and drought when experienced in combination with intensive grazing. Wildrye seed form long, thin awns that extend from the base of each seed unit, which contribute to decreased Pure Live Seed (PLS) percentage in bulk seedlots and make accurate seeding rates difficult to achieve. Current seeding recommendations for southeastern wildrye call for drilling wildrye at 20 lb/acre. The objective of this study was to investigate the effects of seeding rate on biomass yield and population density when drill-planted at 5, 10, 15, and 20 lb/acre PLS. Plots were established at the R.R. Foil Plant Science Research Center in Starkville, MS in the fall of 2012 and 2013. Plant counts were made the following spring of each year, and again that fall. Measured response variables were seedlings/ft^2, mature plants/ft^2, and biomass yield. Data were analyzed using a General Linear Model (GLM) in SAS Statistical Analysis Software. As seeding rate increased, spring plant counts increased correspondingly. There were significant increases in spring-counted plants/ft^2 at each seeding rate. Significant effects on plant populations did not continue into the following fall, with 15 and 20 lb/acre seeding rates having significantly more mature plants/ft^2 than the 5 lb/acre treatment. Seeding rate had no significant effect on biomass yield above 10 lb/acre PLS.

[1]Assistant Research Professor, Department of Plant and Soil Sciences, Mississippi State University, Starkville, MS (jesse.morrison@msstate.edu).

ESTIMATING SOIL ORGANIC CARBON SEQUESTRATION OF PASTURES FROM DEPTH DISTRIBUTION

A.J. Franzluebbers[1]

Sequestration of organic C in agricultural soils is necessary to improve soil health to meet the challenges of climate change, diminishing biodiversity, water quality deterioration, and rising demands for food and fiber production. New assessment approaches are needed to quantify how forage and grazinglands might contribute to soil health improvement and soil organic C (SOC) sequestration. In the southeastern US, forage and grazinglands have been repeatedly shown to stratify SOC concentration with depth. Starting immediately below a zone of tillage influence (i.e. 30-cm depth), SOC concentration is rarely affected by contemporary management due to relatively low C inputs and high decomposition potential, while concentration increases toward the surface in a non-linear manner, presumably with greater inputs of residue and root inputs and changing temperature and moisture conditions. Using these observations as an ecological foundation, SOC sequestration was calculated as the summation of SOC stock greater than the baseline condition at 30-cm depth. Data from literature sources were mathematically fitted with this new approach as a validation. This new approach is unique, but considered valuable to estimate SOC sequestration from a greater diversity of forage and grazingland systems on farms within a region and independent of soil type. An outlook for future research will be posed.

[1]Reseach Ecologist, USDA-Agricultural Research Service, NCSU Campus Box 7620, Raleigh NC 27695 (alan.franzluebbers@usda.gov).

EVALUATION OF HORSE PASTURE SCORE CARD

Krista L. Lea, Reilly Fardy, and S. Ray Smith[1]

With about 1.1 million acres dedicated for equine use in Kentucky, horse pasture health and management are important facets of the equine industry. The UK Horse Pasture Evaluation Program began in 2005 to assist farms by evaluating pasture composition and providing detailed recommendations. This program is currently using 30" by 30" occupancy grids (with 25 - 6" by 6" squares) to determine botanical composition on horse pastures. This method provides accurate and detailed information, but can be time consuming and, in some cases, generates more data than is desired by farm managers. Based on the NRCS Pasture Health Score Card, the UK Horse Pasture Health Score Card was developed in 2019. This method scores pastures on a scale of 1-5 in each of 10 different categories: Desirable Grass Cover, Desirable Grass Diversity, Legumes, Bare Soil and Warm Season Annual Grasses, Broadleaf Weeds, Perennial Weed Grasses, Grazing Management, Stage of Growth, Loafing Areas, and Thatch/Color. In this study, randomly selected pastures were evaluated by the same individual using both the UK Horse Pasture Health Score Card and our traditional occupancy grid. Values from the occupancy grid were then compared to the Horse Pasture Health Score Card for the quantifiable categories. Based on these comparisons, the Score Card shows promise as a reliable, objective, and time saving method to evaluate horse pastures.

[1]Research Analyst, Dept. Plant and Soil Sciences, Univ. of Kentucky, (krista.lea1@uky.edu);
ESM Summer Intern, Dept. Plant and Soil Sciences, Univ. of Kentucky (rmfa227@g.uky.edu);
Extension Professor, Dept. Plant and Soil Sciences, Univ. of Kentucky (Raysmith1@uky.edu).

FALL GRAZING FOR NEWLY WEANED LAMBS

C.H. Gelley[1] and B. J. Campbell[2]

The popularity of grass-fed meats continues to grow across the international and domestic markets. To increase the feasibility of grass-fed systems for Ohio producers, investigation of fall grazing strategies for various species and classes of livestock remains of interest. The objective of this study was to evaluate the effectiveness of three grazing strategies with minimal inputs for sustaining and growing newly weaned lambs in a grazing only system during late fall. Research paddocks were established at the Eastern Agricultural Research Station in Caldwell, OH in August 2019 and grazed during the months of November and December 2019. The forages tested were stockpiled tall fescue pasture, oats, and purple top turnips. Paddocks were a tall fescue and white clover mix prior to a burn down treatment of glyphosate herbicide for the oat and turnip treatments, which were drilled into the paddocks. No supplemental fertilizer was applied during the study. Paddocks were 0.5 acres, strip grazed by six 60-day old lambs averaging 50 pounds each, and replicated three times per forage type. Variables monitored for each treatment included forage quality, dry matter yield, lamb average daily gains (ADG), and lamb health status. Lambs were granted access to new strips of forage every three days, able to back graze, and provided free choice mineral and water for 56 days. Forage quality and dry matter samples and records of lamb weight and health parameters were collected every two weeks. While dry matter yield was not different by forage type in the experiment as a whole ($P > 0.05$), nutritional parameters for digestibility of turnips were preferable when compared to oats and tall fescue ($P < 0.05$).

[1]Agriculture & Natural Resources Educator, Ohio State University Extension, Caldwell, OH, 43724 (gelley.2@osu.edu); [2]Department of Animal Sciences, The Ohio State University, Wooster, Ohio, USA, 44691 (campbell.1279@osu.edu).

FERMENTATION AND AEROBIC STABILITY OF RAIN-DAMAGED ALFALFA/GRASS BALED SILAGES TREATED WITH A PROPIONIC-ACID-BASED PRESERVATIVE

W.K. Coblentz and M.S. Akins[1]

Most new balers are equipped with application systems for preservatives, or the option exists for factory installation at the time of purchase. Within this context, propionic-acid-based products are used commonly for storage of dry hay, but they are not often considered when ensiling baled forages. This study evaluated 4 preservative treatments applied to alfalfa/grass forages (49% alfalfa, 46% grass, 5% broadleaf weeds) that had received 5 mm of unexpected rainfall during wilting. A propionic-acid-based preservative (CropSaver™, New Holland Agriculture, Racine, WI) was applied at rates of 0.01 (control), 0.13, 0.44, or 0.80% of wet bale weight to 24 (4 × 4-ft) round bales made at either 43.6 or 51.6% moisture. Before fermentation, initial forage pH declined linearly ($P < 0.001$) from 6.14 to 5.77 with product-application rate. Conversely, initial forage buffering capacity tended to increase linearly ($P = 0.057$) from 360 to 405 mEq/kg DM. After a 242-day storage period, there were linear increases in concentrations of total alcohols ($P = 0.002$), and particularly 2,3-butanediol ($P < 0.001$), as application rate increased. After storage, plastic wrap was removed from all bales, and they were exposed to air for 11 days. Maximum surface temperatures (6-inch depth) were 132, 94, 73, and 64°F for the 0.01, 0.13, 0.44, and 0.80% rates, respectively, which represented a linear ($P = 0.001$) effect. A similar linear ($P = 0.002$) effect was observed for exposed surface pH (range = 6.52 to 5.41). Use of a propionic-acid-based preservative improved aerobic stability of round-bale silages.

[1]Research Dairy Scientist/Agronomist, U. S. Dairy Forage Research Center, Marshfield, WI (wayne.coblentz@usda.gov); Extension Dairy Nutritionist, Department of Animal and Dairy Sciences, University of Wisconsin, Madison (msakins@wisc.edu).

FORAGE CELL WALL COMPOSITION AND ITS RELATIONSHIP WITH UNDIGESTED NEUTRAL DETERGENT FIBER (uNDF)

Gonzalo Ferreira, Christy L. Teets, and Matias J. Aguerre[1]

Given its filling effect, which may limit feed intake, there is increasing interest in characterizing the undigested neutral detergent fiber (uNDF) of forages. In this study, we determined the uNDF concentration (NDF basis) of leaf and stem tissues of summer annual grasses (corn, sorghum, and pearl millet), alfalfa, and winter annual grasses (barley, oats, rye, ryegrass, and triticale) by incubating samples in the rumen of cannulated cows for a 240-h period. Alfalfa leaves and stems had the greatest uNDF concentration (36.7 and 60.2%, respectively). While the uNDF concentration in leaf blades was similar between summer and winter grasses (20.2%), the uNDF concentration in stem internodes was greater for winter than for summer grasses (45.9 and 39.9%, respectively). The concentration of glucose in the cell wall (CW) does not seem to relate to uNDF, as glucose concentration was lowest for alfalfa tissues (20.1 and 33.1% for leaves and stems, respectively), in the middle for winter grasses (34.3 and 45.0% for leaves and stems, respectively), and greatest for summer grasses (41.5 and 49.3% for leaves and stems, respectively). The ratio of acid detergent lignin (ADL) to neutral detergent fiber (NDF) was greatest for alfalfa tissues (19.3 and 22.5% for leaves and stems, respectively) and least for summer grasses (6.8 and 5.0% for leaves and stems, respectively) and winter grasses (6.9 and 7.1% for leaves and stems, respectively). ADL/NDF ratio seems to explain uNDF when comparing forage types (legumes vs. grasses) but not when comparing grasses.

[1]Dairy Management Extension Specialist, Department of Dairy Science Research, Virginia Tech, Blacksburg, VA (gonf@vt.edu); Laboratory Research Assistant, Department of Dairy Science Research, Virginia Tech, Blacksburg, VA (cteets77@vt.edu); Ruminant Nutritionist, Animal and Veterinary Sciences Department, Clemson University (maguerr@clemson.edu).

FORAGE MASS AND STRUCTURAL RESPONSES OF STOCKPILED ALFALFA-BERMUDAGRASS SYSTEMS

L.S. Silva, J. Burt, J.J. Tucker, M.K. Mullenix, K. Groce, L. Jacobs, and K. Manson[1]

Incorporation of alfalfa into bermudagrass systems can increase forage production and nutritive value, while adding nitrogen through biological fixation. Although this system can extend forage production season, climatic conditions often affect growth, harvest and forage nutritive value. The objective of this study was to determine forage mass, botanical composition, and structural responses of alfalfa-bermudagrass systems managed under five stockpiling strategies (6-, 8-, 10-, 12- and 14-weeks). A randomized complete block design with eight replications was used. Harvested forage samples were manually separated into alfalfa, bermudagrass and weeds components. After, sample bags were dried at 140 F to constant weight and weighed to determine forage mass. Additional measurements corresponded to ground cover, light interception characteristics, and alfalfa stand density, persistence, lodging and leaf shattering ratings. Alfalfa corresponded to over 40% of canopy composition for all treatments. Increased regrowth period contributed to increased alfalfa lodging and leaf shattering, and bare soil percentages. Canopy leaf area index and forage digestibility decreased with increased regrowth period. These preliminary results supported viability of this mixture for use in the region; however, longer regrowth periods may affect nutritive value due to increased maturity and decrease in leaf percentage.

[1]Post-Doctoral Research Fellow, Auburn University, Auburn, AL (lzs0105@auburn.edu); Graduate Student, University of Georgia, Tifton, (justin.burt@uga.edu); Assistant Professor, University of Georgia; Assistant Professor, Auburn University, Auburn, AL (mullemk@auburn.edu); Graduate Student, Auburn University, Auburn, (jlj0049@auburn.edu); Graduate Student, Auburn University, Auburn, (kmb0108@auburn.edu).

FORAGE QUALITY OF MIXED SYSTEMS IN ORGANIC DAIRY PRODUCTION

S. R. Smith, K. Payne, K. Lea, D. Butler, G. Bates and G. Pigetti[1]

The University of Tennessee and the University of Kentucky are completing a five year USDA-NIFA-OREI grant referred to as the Southeast Organic Dairy Project. The overall project objective was to develop practical, research-based recommendations for organic forage management to maximize forage and herd productivity, herd health, and herd profitability in the southeastern U.S. This poster highlights the forage quality (CP, NDF and ADF) of warm and cool season forage mixtures using small plot trials conducted in KY and TN. The nine mixtures and monoculture controls included: 1 - red clover (RC); 2 - orchardgrass (OR); 3 - RC + OR; 4 - alfalfa (A) + OR; 5 - RC + OR + tall fescue (TF); 6 – RC + OR + crabgrass (CG); 7 - RC + annual ryegrass (AR) + CG; 8 – crimson clover (CC) + AR + sorghum sudangrass (SS) + cowpea (CP); and 9 – CC + AR + CG + annual lespedeza (AL). Mixtures with cool season perennials and annuals consistently maintained higher forage quality over harvests and years than warm season annuals. Not surprisingly, the presence of legumes increased forage quality across all mixtures. In a separate analysis, the perennial cool season mixtures provided the most cost efficient forage production and forage quality in comparison to the annual cool/warm season mixtures because of the cost of annual planting. This research demonstrated that complex forage mixtures can be beneficial for organic dairy producers, but perennial cool season mixtures likely provide the most economical forage production.

[1]Professor, Ext. Specialist, Univ. of Kentucky, Lexington, KY (raysmith1@uky.edu); Applied Forage Systems Specialist, Virginia Tech So. Piedmont Agric. Center, Blackstone, VA (Kathrynpayne@vt.edu); Research Analyst, Univ. of KY, Lexington, KY (krista.lea1@uky.edu); Associate Professor, Univ. of TN, Knoxville, TN (dbutle14@utk.edu); Professor, Ext. Specialist, Univ. of TN, Knoxville, TN (gbates@utk.edu); Associate Professor, Univ. of TN, Knoxville, TN (pighetti@utk.edu).

FORAGE QUALITY: FACTORS AFFECTING, TESTING, AND ANIMAL PERFORMANCE

J. A. Jennings[1]

Many factors affect forage quality such as temperature and soil fertility. Nitrogen fertilizer can impact forage crude protein content to an extent. Cool-season forages tend to be more digestible than warm-season forages. But, nearly all forages can be considered to have low quality when they become mature. As plants mature, highly digestible components of plant cells shrink in percentage of the forage mass. While it is true that specific forages are generally higher in forage quality than others, most forages have adequate quality to sustain livestock production when managed to be in a vegetative growth phase. Most forages can be high in forage quality when vegetative. Few retain high quality in later stages of maturity. Laboratory test results usually report forage quality factors such as crude protein and energy. Other values are often reported such as Relative Feed Value (RFV) and Relative Feed Quality (RFQ). Those are single number factors used to rank forages based on a set quality standard value. Unfortunately, they cannot be used to develop rations for feeding livestock. The level of quality desired depends upon the animal being fed and the level of output needed from that animal. Pasture quality varies during the growing season and hay quality becomes a fixed value once it is harvested. Producers need to be aware of the quality requirements needed for their livestock and manage their forage resources to attain that level.

[1]Professor, Forage Specialist, University of Arkansas Division of Agriculture, Little Rock, AR (jjennings@uaex.edu).

FORAGE YIELD AND NUTRITIVE VALUES OF FIVE SOYBEAN CULTIVARS

Thompson, S.[1], Jacobson, A.[2], Carrell, R.[1], Cole, M.[1], Dillard, L.[1]

Soybeans can be useful a useful warm-season legume option. The release of new cultivars has renewed interested in soybeans as a forage crop. The current study evaluated 'Tower of Leaves', 'Stonewall', 'Laredo', 'AG79X9', and 'AG64X8' (T, S, L, A and X, respectively) grown in the 2019 and 2020 growing seasons. These cultivars (C) were evaluated at two row spacings (36 and 71 cm), two locations (E.V. Smith Center in Shorter, AL, and Wiregrass Center in Headland, AL), and three harvest heights (HH, 10, 15, and 20 cm). Whole plots were harvested using a forage harvester for yield determinations, and grab samples were taken for forage quality analysis. Grab samples were dried in a 55°C forced air oven and then ground to 1 mm. Forage quality was assessed using NIRS. This study was conducted using a randomized, complete block design and data were analyzed using PROC Glimmix in SAS (SAS Inst., Cary, NC). A significant year × HH ($P < 0.0001$) and year × C ($P < 0.0001$) effect was observed in forage yield. Greater yields were observed from S and T, compared with L (2,074, 1,978, and 1,348 kg/ha, respectively). In 2019, neutral detergent fiber (NDF) and acid detergent fiber (ADF) were affected ($P < 0.01$) by both HH and C, while crude protein (CP) was only affected ($P < 0.0001$) by HH. The lowest NDF and ADF was observed in S (30.84 and 29.27%, respectively; ($P < 0.01$). The cultivar S shows the most promise for use in grazing systems, as it has greater yield and forage quality compared to other cultivars.

[1]Dept. of Animal Sciences; Auburn University; [2] Dept. of Crop, Soils, and Environmental Science; Auburn University.

GRAZING COVER CROPS TO EXTEND PRODUCTION ON WINTER-VACANT CROPLAND

J. Cox-O'Neill, D. Harmon, M. Poore, A. Franzluebbers, J. Eisemann, and C. Pickworth[1]

Continuously evolving factors in agriculture, such as fluctuating commodity prices, shifting producer demographics, and widespread concern for the environment has piqued interest for more diverse agricultural systems. Forage production, soil characteristics, and subsequent corn yield was evaluated with and without grazing stocker steers (653±56 lb; at stocking rate of 1.5 steer/acre) on rye monoculture or rye-turnip mixture as cover crop treatments, at three Coastal Plains locations in North Carolina for two years. Pre-grazing forage yield was greater ($P<0.01$) for rye-turnip mixture (1685 and 3996 lb DM/acre, Year 1 and 2, respectively) than rye monoculture (1399 and 1376 lb DM/acre, Year 1 and 2, respectively). Soil-test biological activity tended to be greater ($P=0.10$) in grazed areas than non-grazed areas in Year 1 (297 vs. 275 mg/kg, respectively), but not different ($P=0.85$) in Year 2 at 0-2 inch depth. Corn grain yield did not differ ($P=0.78$) between forage and grazing treatments in Year 1 (97 bu/acre), but had a tendency to be greater ($P=0.06$) for grazed (155 bu/acre) than non-grazed areas (132 bu/acre) in Year 2. Steer average daily gain did not differ ($P=0.77$) between forage treatments (2.33 lb/d). Grazing either forage treatment resulted in similar steer gains and positive impacts on soil-biological activity and subsequent corn yield at this stocking rate. Integrated crop-livestock systems may enhance livestock and crop production and provide environmental benefits.

[1]Ph.D. Candidate, NCSU (jlcox8@ncsu.edu); Extension Livestock Specialist, NCSU (ddharmon@ncsu.edu); Ruminant Nutrition Specialist, NCSU (mhpoore@ncsu.edu); Research Ecologist, USDA-ARS (alan.franzluebbers@usda.gov); Professor, NCSU (joan_eisemann@ncsu.edu); Assistant Professor, NCSU (clpickwo@ncsu.edu).

IMPACT OF CATTLE REMOVAL DATE ON ANIMAL AND COVER CROP PERFORMANCE

R.C. Carrell[1], S.L. Dillard[1,2], M.K. Mullenix[1], A.V. Gamble[2], and R.B. Muntifering[1]

Cool-season cover crops have been shown to provide environmental benefits for land in continuous row crop production. However, evidence is still needed to verify the effects of cattle grazing on cover crop and cash crop production throughout an extended time period. Our objective was to determine animal and forage performance when grazing a cool-season annual cover-crop mixture in a peanut-cotton rotation. Twelve 3-acre pastures were established in a mixture consisting of 'Cosaque' oats (*Avena strigose)*, 'FL401' rye (*Secale cereal)*, 'AU Sunrise' crimson clover (*Trifolium incarnatum)*, and 'T-raptor' brassica (*Brassica napus* × *B. rapa*) and randomly allocated to be grazed until either February, March, or April with a negative control of no grazing. Three tester steers were randomly placed in each paddock with the exception of control paddocks and allowed free choice grazing. Animals were weighed every 30 d for determination of average daily gain (ADG). Forage was harvested bi-weekly and analyzed for yield, NDF, ADF, and CP. Forage yield showed differences between Control and February (4,063 vs. 1,857 lb/acre; $P = 0.002$), Control and March (4,063 vs. 2,005 lb/acre; $P < 0.001$), and Control and April (4,063 vs. 2,173 lb/acre; $P < 0.001$). No differences were detected in ADG (2.72 lb/day, $P = 0.56$), NDF (48.57%; $P = 0.78$), ADF (29.72%; $P = 0.92$), or CP (12.86; $P = 0.74$) among treatments. These results indicate that although grazing length can have an impact on forage yield, there is no effect on plant nutrient composition or individual animal performance when grazing cover crops.

[1]Department of Animal Science, Auburn University, Auburn, rcc0057@auburn.edu;
[2]Department of Crop, Soil, and Environmental Sciences, Auburn University, Auburn.

IMPACT OF DEFOLIATION HEIGHT ON CROWN AND SOIL TEMPERATURE

Jessica Buckman, Garrett Hatfield, and Chris Teutsch[1]

Defoliation management in pastures and hayfields can impact the productivity and persistence of cool-season grasses. Past research has shown that defoliation height and frequency impacts both leaf area and energy reserves in the plant. However, limited data is available on the impact of defoliation on the microclimate in cool-season grass stands. The objective of this study was to evaluate the impact of defoliation height and frequency on soil and crown temperatures of an established cool season grass stand. In the summer of 2019 and 2020, a study was conducted at the University of Kentucky's Research and Education Center, located near Princeton, KY. The experimental design was a randomized complete block with four replications. An established tall fescue (*Schedonorus arundinaceus* (Schreb.) Dumort., nom. cons.) stand was defoliated to 1.0 and 4.5-inches weekly and monthly to 4.5 inches. Crown and soil temperature were measured at 15-minute intervals using HOBO MX2303 loggers (Onset Computer Corporation, Bourne, MA). The 1.0-inch defoliation height increased crown temperature by more than 10 degrees Fahrenheit on the days with air temperature that exceeded 90 degrees Fahrenheit. Soil temperature was also higher at the 1-inch defoliation height, but to a lesser extent. Our data, indicate that maintaining adequate residual leaf area in pastures and hayfields (>4.5-inches) could moderate the microclimate at and below the soil surface, which may in turn aid in the persistence of cool-season grass stands in the transition zone states like Kentucky.

[1]Undergraduate Summer Intern, Hutson College of Agriculture, Murray State University, Murray, KY 42071; Undergraduate, Dept. of Animal Science, Iowa State University, Ames; and Associate Extension Professor, Dept. of Plant and Soil Sciences, Univ. of Kentucky, Lexington (chris.teutsch@uky.edu).

IMPACT OF DRY MATTER AT ENSILING ON THE PH AND NUTRIVE VALUE OF BALEAGE

A.E. Teutsch and C.D. Teutsch[1]

Harvesting hay as baleage can allow for more timely harvest, especially in spring months when curing conditions are not ideal. Forage conserved as baleage undergoes anaerobic fermentation in which the sugars are converted to lactic acid, lowering final pH. To successfully conserve forage as dry hay, moisture concentrations must be less 18%. In contrast, the ideal moisture concentration for baleage is between 55 and 65%. There are a number of producers harvesting and wrapping hay that is not quite "dry enough" to bale, but is too dry to ensile. This produces a product commonly referred to by producers as "sweet hay". The objective of this study was to evaluate the impact of dry matter at ensiling on the on final pH and nutritive value of cool-season grasses. This study was conducted at UK Research and Education Center located near Princeton, KY. Two mixed cool-season grass hay fields were sampled every two hours during daylight starting at cutting and ending when the hay was dry enough to bale. Each sample was immediately chopped into 2 to 3-inch lengths and sealed in a vacuum bag (3/4 full). After 6 weeks of fermentation, the vacuum bags were sampled. Final pH increased as DM at ensiling increased. Dry matter at ensiling had little impact on nutritive value. Our data indicate that if oxygen is excluded from the forage, ensiling in the range of 18 to 55% moisture may be a viable range management tool.

[1]Undergraduate Summer Intern, University of Kentucky, Lexington, KY 40546 (AETeutsch@uky.edu); and Extension Associate Professor, University of Kentucky Grain and Forage Center of Excellence, Princeton, KY 424454.

IMPACT OF FERTILIZER TYPE, SEEDING COATING, AND DURATION OF EXPOSURE ON THE GERMINATION OF ORCHARDGRASS SEED

L.D. Sparks, C.D. Teutsch, and E.L. Ritchey[1]

Cool-season grasses such as orchardgrass, are important components of forage systems in transition zone states like Kentucky. In order to maintain dense and vigorous sods, improved cool-season grass varieties are sometimes overseeded into existing stands in late-winter or early-spring. In many cases seed is mixed with fertilizer and top-dressed onto pastures. Little data are available on the impact of fertilizer type or duration of exposure on the germination of raw and coated grass seed. The objective of this study was to evaluate the impact of two fertilizer types, muriate of potash and a blended fertilizer (urea, diammonium phosphate, and muriate of potash), and the duration of exposure (0 to 28 days) on the germination of an improved orchardgrass variety that was raw or coated. Mixing seed with the blended fertilizer resulted in a quadratic decease in germination rate for the raw seed and coated seed. This rate of decrease was greater for the coated seed. After 14 days of exposure to the blended fertilizer, the germination rate of the coated and raw seed was 0 and 17%, respectively. Combining seed with muriate of potash resulted in a linear decline in germination with the decline being similar for both the coated and raw seed. Overall, the rate of decrease was considerably less than that of the blended fertilizer. Results of this study indicate that the combination of seed coating and blended fertilizer had the most detrimental impact on orchardgrass germination.

[1]Undergraduate Summer Intern at UKREC, University of Tennessee at Martin, TN 38238; and Extension Associate Professors, University of Kentucky Grain and Forage Center of Excellence, Princeton, KY 424454 (Chris.Teutsch@uky.edu).

IMPACT OF FERTILIZER TYPE, SEEDING COATING, AND DURATION OF EXPOSURE ON THE GERMINATION OF RED CLOVER SEED

L.D. Sparks, C.D. Teutsch, and E.L. Ritchey[1]

Legumes are important components in grassland ecosystems. Red clover is one the most used legumes in the transition zones states like Kentucky. In order to maintain legumes in grass pastures, improved red clover varieties are often overseeded in the late-winter or early-spring. In many cases seed is mixed with fertilizer and top-dressed onto pastures. Little data are available on the impact of fertilizer type or duration of exposure on the germination of raw and coated red clover seed. The objective of this study was to evaluate the impact of two fertilizer types, muriate of potash and a blended fertilizer (urea, diammonium phosphate, and muriate of potash), and the duration of exposure (0 to 28 days) on the germination of an improved red clover variety that was raw or coated. Mixing seed with the blended fertilizer resulted in a linear decease in germination rate for the raw seed and quadratic decrease for the coated seed. After 20 days of exposure to the blended fertilizer, the germination rate of the coated and raw seed was 0 and 60%, respectively. Combining seed with muriate of potash resulted in a linear decline in germination rate with the decline being greater for the coated seed. Overall, the rate of decrease was considerably less than that of the blended fertilizer. Results of this study indicate that seeding coating enhanced the detrimental effects of fertilizer on seed germination.

[1]Undergraduate Summer Intern at UKREC, University of Tennessee at Martin, TN 38238; and Extension Associate Professors, University of Kentucky Grain and Forage Center of Excellence, Princeton, KY 424454 (Chris.Teutsch@uky.edu).

INCREASING GRASSLAND KNOWLEDGE IN WV WITH PRESCRIBED GRAZING PROFESSIONAL DEVELOPMENT PROJECT

B.M. Goff, E.B. Rayburn, T.J. Basden[1], and J.I. Wolford[2]

A series of prescribed grazing workshops was developed through collaboration with the WVU Extension Service, NRCS, and the WV Conservation Agency to train personnel in implementing grazing management to better suit the need of local landowners and other clientele. A total of 12 workshops covering six topic areas (vegetation management, animal management, conservation planning, pasture assessment, economics, and soil fertility/ management), each consisting of a 2-3 day program that containing a mixture of in-classroom lectures and hands-on training in the field. At the end of the workshop, a total of 38 participants became Certified Forage and Grassland Professionals (CFGP) and passed with an average score of 93%, which was an improvement of 33% compared to a pre-test given at the initial workshop.

[1]ANR Extension Agent-Mason/Putnam Counties, Extension Forage Specialist, and Extension Nutrient Management Specialist respectively, West Virginia University Extension Service.
[2]Mangager/Agronomist, USDA-Plant Material Center, Alderson, WV.

INTEGRATED CROP-LIVESTOCK SYSTEMS IN THE SOUTHEASTERN USA

J.B. Rushing[1], J.C. Lyles[2], B.S. Bass[3], and K.R. Waddell[3]

Integrated crop-livestock systems (ICLS) are designed to produce both cash crops and livestock on the same land base in order to reduce nutrient concentration and environmental risk, improve soil structure and function, exploit natural pest control processes, and increase overall farm productivity and economic sustainability. Two experiments are currently being conducted at the Coastal Plain Branch Experiment Station in Newton, MS. Experiment I consists of three cover crop treatments followed by no-till soybean (*Glycine max* L.) production. Cover crop treatments are grazed using weaned commercial beef (*Bos* sp.) steers at an initial stocking rate of 1,000 lb/a. In Experiment II, twelve cover crop treatments combined with two tillage treatments are planted following corn (*Zea mays* L.). Bred beef heifers are used to graze cover crop treatments and replicate compaction and nutrient cycling. In both experiments, soil health measurements (compaction, bulk density, nutrient availability, moisture, organic matter, and total N) are collected. Also, crop (moisture and test weight), cover crop (dry matter and nutritive value), and beef (ADG - average daily gain; GAIN - gain per acre) yields are compared across treatments. First year data analysis shows no differences in soil compaction between grazed and un-grazed plots for both studies. Nutritive value of cover crops is more than adequate to meet nutritional demands of grazing livestock within each respective study. No differences were observed in ADG or GAIN in Experiment I. A second year of data collection and analysis is currently underway, including comprehensive economic comparisons within each experiment.

[1]Associate Research/Extension Professor; Coastal Plain Branch Experiment Station, Newton, MS; [2]Research Associate I; Coastal Plain Branch Experiment Station, Newton, MS; [3]Graduate Research Assistant; Coastal Plain Branch Experiment Station; Newton, MS.

IS THERE AN ECONOMIC ADVANTAGE TO PLANTING DIVERSE SUMMER ANNUAL FORAGE MIXTURES?

K. M. Mercier, C. D. Teutsch, S. Ray Smith, K. H. Burdine, E. S. Vanzant, and E. L. Ritchey[1]

Biodiversity in perennial systems often results in increased productivity, in part due to N sharing of legumes. This experiment was conducted to determine if this relationship could be found in a summer annual forage system. Treatments included a grass monoculture, a simple 3-species grass-legume mixture, and a complex 11 species grass-legume-forb mixture. Nitrogen rates from 0-200 lb N/A were applied to determine if mixtures required less N to obtain similar yields. All treatments responded to N similarly, regardless of legume inclusion. Nitrogen rate linearly increased yields in three out of four environments by an average of 15 lb DM/lb N. In one out of four environments, yields were greater in the mixtures as compared to the monoculture ($p < 001$), but yields were similar for all forage treatments in the remaining three environments ($p > 0.06$). Complex mixtures resulted in \$15 greater seed costs compared to the monoculture with no yield advantage in 3 out of 4 environments due to poor competitive ability of dicotyledonous species. Legume proportions were < 5% in both mixtures and may have been one reason no N rate x mixture interaction was observed. Since the legume proportion was so low in this study, it is not possible to make any meaningful N recommendations for summer annual grass-legume mixtures. However, the results of this study do indicate that it is important to plant species with compatible morphology and growth rates to reduce competition in mixtures.

[1]Graduate Student, University of Kentucky Research and Education Center at Princeton (UKREC), Princeton, KY (kelly.mercier@uky.edu); Extension Associate Professor, UKREC, Princeton, KY (chris.teutsch@uky.edu); Extension Professor, University of Kentucky (UK), Lexington, KY (raysmith1@uky.edu); Associate Extension Professor, UK, Lexington, KY (kburdine@uky.edu); Associate Professor, UK, Lexington, KY (evanzant@uky.edu); Extension Associate Professor, UKREC, Princeton, KY (elritc2@uky.edu).

MAKING HAY

Lisa Baxter, Marcelo Wallau, Leanne Dillard, Uttam Saha, Chip Blalock, and Ash Alt

Hay is the most important source of stored forage for beef cattle for supplementary winter feeding. While good hay can result in improved animal performance, poor quality hay reduces gain, increases waste, and is ultimately more expensive. Unpredictable weather and the humid climate can make hay production a challenge in the Southeast U.S. Poor maintenance and improper management practices can further reduce product quality. This session will cover the foundation of good hay production - management practices, techniques, and technologies which can improve hay production in the Southeast. Good hay starts with sound harvest management. The first presentation will focus on harvesting timing, showcasing real-life examples to illustrate the importance of aiming for quality instead of quantity. However, quality forage can be ruined by improper use of hay equipment. The second presentation will show how to properly adjust equipment commonly used in the hay field, and compare differences made to each tool and its impact on forage quality. Technology for agricultural production has also increased rapidly in the past decade, from improved conditioners to inoculants, and sensors to predict production and quality. The third presentation will highlight some of the simple tools producers can implement in the hay operations. Finally, join us in the 2020 Southeastern Hay Contest award ceremony, featuring top hay producers on a collaborative outreach effort from 13 Southeastern states.

Forage Extension Specialist, Crops and Soils Science Dept., University of Georgia, Tifton, GA (baxterl@uga.edu); Forage Extension Specialist, Agronomy Dept. University of Florida, Gainesville, FL (mwallau@ufl.edu); Forage Extension Specialist, Animal Science Dept., Auburn University, Auburn, AL (dillasa@auburn.edu); Program Coordinator, Agricultural and Environmental Services Lab, University of Georgia, Athens, GA (sahau@uga.edu); Executive Director, Sunbelt Agricultural Exposition, Moultrie, GA (chip@sunbeltexpo.com); Field Execution Manager, AGCO Corp., Duluth, GA (ash.alt@agcocorp.com)

MAKING ROUND BALE SILAGE – LESSONS LEARNED FROM AN ON-FARM SURVEY OF BALEAGE IN KENTUCKY

Jimmy Henning, Jeff Lehmkuhler, Brandon Sears, Corinne Belton, Levi Berg, Don Sorrell, and April Wilhoit[1]

Abstract

The ability to make silage in round bale packages allows producers to avoid rain damage and produce high quality stored forage. The proliferation of bale wrappers (both individual and inline types) has made this technology more available to producers. Although the process for making baleage is well documented, producers have experienced poor fermentation and in some cases botulism toxicity from baleage. The parameters of good silage are well known and include pH below 5.0 and lactic acid concentrations above 3% on a DM basis. However, tests of farmer-produced baleage reveals that often these target values are not achieved. To better understand the fermentation characteristics of Kentucky baleage, a survey was conducted of round bale silage samples from several Kentucky counties in 2017, 2018 and 2019. Moisture content (MC) at baling was the greatest determinant of fermentation success as measured by pH and lactic acid concentration. Moisture contents were highly variable, with more outside the recommended range of MC (40 to 60%) than within. Farmer practices including wilting time and equipment used were recorded for the 2019 samples. All of the well-accepted practices for making baleage were confirmed in this survey (cutting on time, wilting to proper MC, dense bales, achieving and maintaining anaerobic storage conditions). Baleage samples exceeding 65% MC had elevated butyric acid concentrations, indicating secondary fermentation by clostridial bacteria.

Introduction: Baleage is the process of making silage in bales of wilted forage which is wrapped in plastic to create an anaerobic environment. Baleage allows more timely harvest of forage crops, producing consistently higher quality stored feed than haying which requires longer periods of good curing weather. Key to this process is wilting the forage sufficiently to achieve the recommended moisture content (MC) leading to adequate fermentation within the wrapped bales. Good fermentation depends on achieving anaerobic conditions and rapid fermentation of available soluble carbohydrates to quickly lower bale pH by the production of lactic and acetic acid. The accumulating acids lower the pH, ideally to pH 5.0 or below (slightly higher in legumes). This acidity prevents spoilage during storage and when re-exposed to air at feeding. Well-fermented baleage is generally accepted to have more than 3.0% lactic acid and less than 0.1% butyric acid (dry matter basis).

---

[1]Extension Professor (jimmy.henning@uky.edu); Associate Extension Professor (jeff.lehmkuhler@uky.edu), Extension County Agents for Agriculture and Natural Resources (brandon.sears@uky.edu, Corinne.belton@uky.edu, levi.berg@uky.edu, dsorrell@uky.edu, april.wilhoit@uky.edu, University of Kentucky, Lexington, KY 40546.

Making Good Baleage: The recommended practices for making good baleage are generally consistent across all regions of the US. Forage should be cut at an early stage of maturity so there are adequate soluble carbohydrates for the bacteria on the forage to ferment. Forages should be wilted to between 40 and 60% MC and preferably in the upper half of this range (50 to 60% MC). The time required for wilting will range from hours to days depending on the time of year, type of crop and thickness of the windrow. Forage should be baled into dense, uniform bales and wrapped with six layers of UV-resistant, stretch wrap plastic within 24 hours of baling. Although research has shown that four layers is adequate for ensiling in round bales, six layers gives an added level of safety and protection against punctures from forage stems and from exterior damage (Hancock and Collins, 2006). Since the plastic is somewhat (but slowly) permeable to oxygen, six layers will also give more protection from oxygen diffusion into the bale. Bales should be monitored for punctures; any damage must be repaired quickly with adhesive tape designed specifically for baleage. The ensiling process will take four to six weeks.

As simple as this process sounds, baleage can and does go bad - that is fail to ferment properly. With baleage ensiled too dry, fermentation is incomplete and good storage relies on achieving and maintaining anaerobic conditions (Shinners et al., 2009). In the case of high moisture baleage (70% and above) and especially when forages are mature (lack sufficient fermentable carbohydrate (CHO), bales do not generate sufficient lactic acid to prevent a detrimental secondary fermentation by Clostridium bacteria. These bacteria metabolize available carbohydrates (including lactic acid) and protein and generally produce butyric acid, acetic acid, ammonia and sometimes other toxins (botulinum toxin for example) depending on the species of Clostridium present. Easily the most problematic of these bacterial types is *Clostridium botulinum*, the species responsible for the production of the botulism toxin.

Ordinarily present in low numbers on forage, Clostridium bacteria are introduced into baleage most often from soil or applied manure. Soil is most commonly introduced into baleage from aggressive raking. In addition, driving rain can splash soil onto swaths and windrows. Feeding poorly fermented, high moisture baleage can be costly. Baleage that has undergone clostridial fermentation can have effects ranging from lowered intake to introduction of intestinal toxins to botulism. Nearly every year, cattle are lost to botulism toxicity from poorly fermented baleage. Identifying problem baleage and preventing cattle losses would be highly beneficial.

A survey of Kentucky baleage was begun in 2017 to determine the fermentation characteristics in various lots of baleage. Samples were collected on selected lots of baleage in Kentucky and analyzed for forage quality, pH and fermentation profiles in a commercial, certified laboratory. In 2019, additional data was collected on general production practices, such as equipment type and the interval between cutting and baling. All lots were allowed to ferment for at least four weeks before sampling. Fermentation results were similar for 2017, 2018 and 2019, so only 2019 data is presented.

Survey Results: Baleage moisture content ranged from 22 to 79%, with more samples above 60% MC (18) than below 40% (12) (Figure 1). More samples were outside of the recommended MC range (40 to 60%) than within (30 vs 27). Moisture content at baling was the major factor impacting lactic acid and butyric acid production (Figures 2 and 3). Only when MC approached 60% did samples in this survey achieve 3% lactic acid and a pH of 5.0, generally accepted threshold values for good fermentation. Fermentation characteristics remained good for most samples at MC between 60 and 70, but the results became less predictable. Above 70% MC, butyric acid production was common. Lactic acid values generally declined at MC above 70%, presumably due to its metabolism by Clostridium bacteria.

Producer Practices: Most producers had some type of silage baler or baler adapted to handle silage. All used wheel rakes, but mowers were a mix of conditioning and non-conditioning types. Wilting time was difficult to summarize but varied from hours to a day or more depending on conditions. All producers were using at least four layers of plastic, and most six or more. All of the high butyric samples were small grains, and were high in moisture. None of the producers used any form of silage or baleage inoculant.

Moisture Content and Baleage Fermentation: Moisture levels in Kentucky baleage in 2019 were highly variable (Figure 1). Moisture content at baling was related to the production of lactic acid and lowering bale pH. Lactic acid values tended to rise with MC up to 60 to 65%, and then decline. In these samples, baleage needed to be 55% MC or above to consistently produce adequate lactic acid. All samples at or below 60% MC had acceptable butyric acid levels but one (<0.1%). These results are consistent with other reports of limited fermentation in baleage, especially at less than optimum MC (Muck, 2006).

Fermentation reports showing low lactic acid and pH >5.0 cause concern with producers and uncertainty among extension and industry consultants regarding interpretation. Shinners et al. (2009) found that baleage at or below 60% MC and with less than target lactic acid numbers should be stable in storage and feeding, if anaerobic conditions are maintained until feeding and bales are consumed within one week. Therefore, baleage with lower MC and lactic acid should be stable in storage so long as plastic integrity is maintained. These bales will not be very stable after exposure to air and should be fed quickly.

Baleage with MC between 60 and 70% produced high quantities of lactic acid (1.9 to 5.8% DM basis) but also above optimal butyric acid (>0.1%) in 10 of 14 samples. Above 70% MC, all samples had problematic levels of butyric acid. Further, butyric acid was present (0.23 to 6.0%) in every lot of small grain baleage above 60% MC. Type of crop/season of production should also be considered when determining risk; small grain baleage is produced in early spring when lactic acid bacteria numbers and lower temperatures likely limit beneficial fermentation. Poor curing conditions and wet soils make it difficult to wilt forage to the desired moisture levels. Some producers in this Kentucky database wilted small grain forage for 72 hours and still did not achieve moisture contents below 70%. Baleage inoculants would likely improve fermentation in spring produced baleage, but was not used on any lots tested in any year.

What About Ash?: Ash is the mineral content of forage, and should be between six and eight percent of DM unless soil contamination has occurred. Elevated ash levels (above 10 to 11% DM basis) are a concern since dirt is the primary way that Clostridium bacteria are introduced into baleage. Ash content ranged from 7 to 15% in the 2019 samples. There was no clear relationship between elevated ash content and high butyric acid. However, there was evidence that the ash effect on clostridial fermentation might be seasonal. In two samples with ash levels above 14%, small grain baleage had greater butyric acid than sorghum-sudangrass baleage produced later that season from the same field (0.44 vs 0.01% DM basis). High ash content alone is not predictive of clostridial fermentation, but should be considered as a risk factor.

Summary: All of the well-accepted practices for making baleage were confirmed in this survey (cutting on time, wilting to proper MC, dense bales, achieving and maintaining anaerobic storage conditions). Baleage testing for quality and fermentation is highly recommended. Normal forage testing will not measure actual botulism toxin formation, but it will reveal risk factors that make it more likely. These risk factors are high MC (above 65%), pH > 5.0, ash content > 15% and high butyric acid production. To be clear, the presence of butyric acid above 0.1% is not proof of botulism or even a perfect predictor of poor performance. But high levels of butyric acid should be considered as increasing the risk of feeding problems, including the possibility of botulism. With a comprehensive forage test that includes a fermentation profile and measurement of the ash content, feeding problems due to poor fermentation become preventable or at least more predictable.

Literature Cited:

Hancock, D.W. and M. Collins. 2006. Forage preservation method influences alfalfa nutritive value and feeding characteristics. Crop Science 46(2):688-694.

Muck, R.E. 2006. Fermentation characteristics of round-bale silages. In Proceedings 60th Southern Pasture and Forage Crop Improvement Conference, April 11-13, 2006, Auburn University. https://agrilife.org/spfcic/files/2013/02/muck.pdf

Shinners K.J., B.M. Huenink, R.E. Muck, and K.A Albrecht. 2009. Storage characteristics of large round and square alfalfa bales: Low-moisture wrapped bales. Trans. ASABE. 2009; 52: 401.

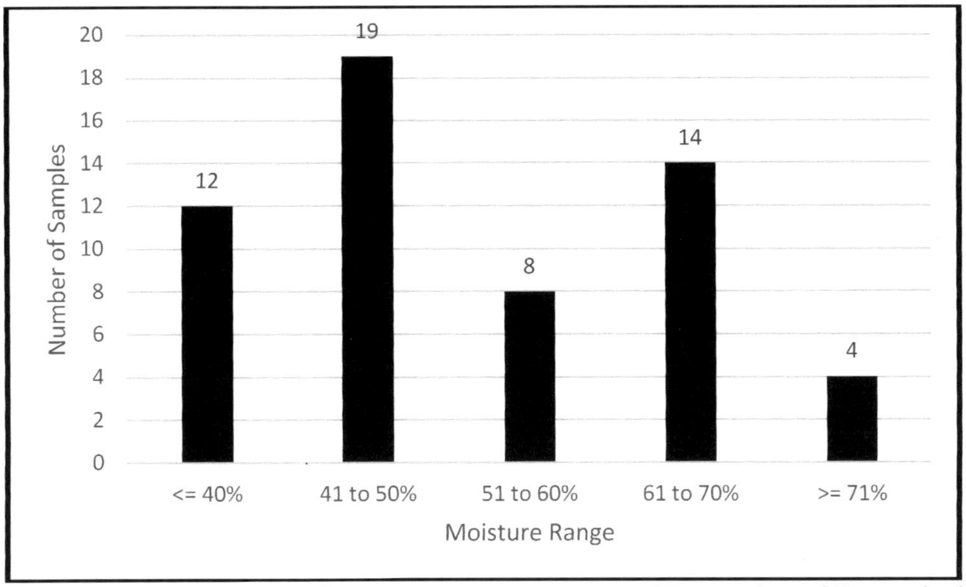

Figure 1. Distribution of moisture content among 57 samples of Kentucky baleage in 2019.

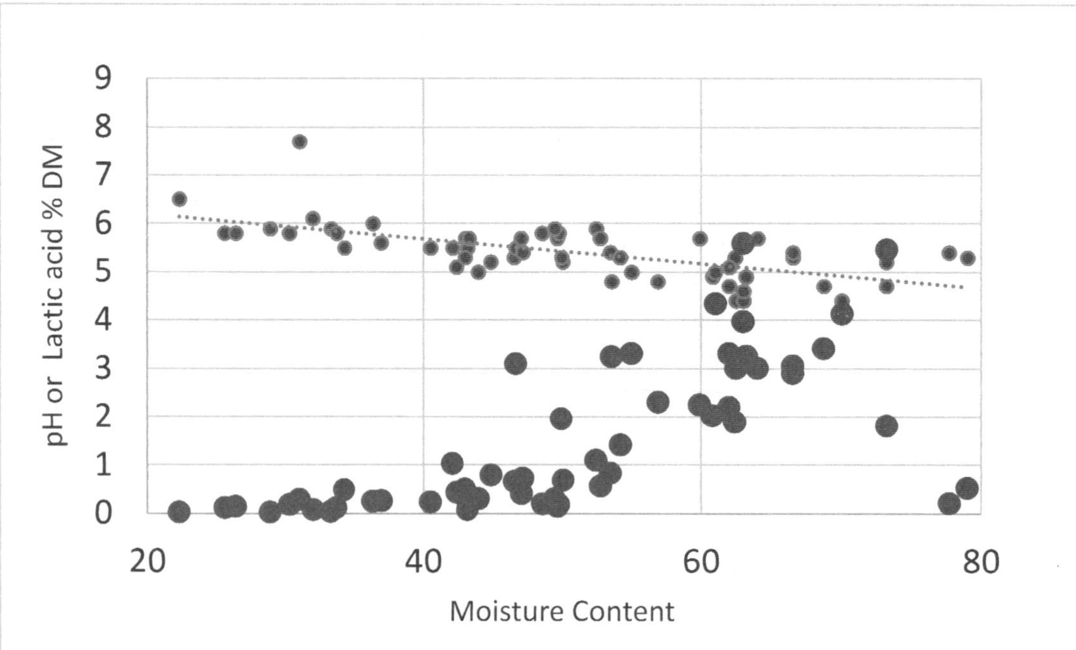

Figure 2. The effect of moisture content on lactic acid (large circles) and pH (small circles) of 2019 Kentucky baleage.

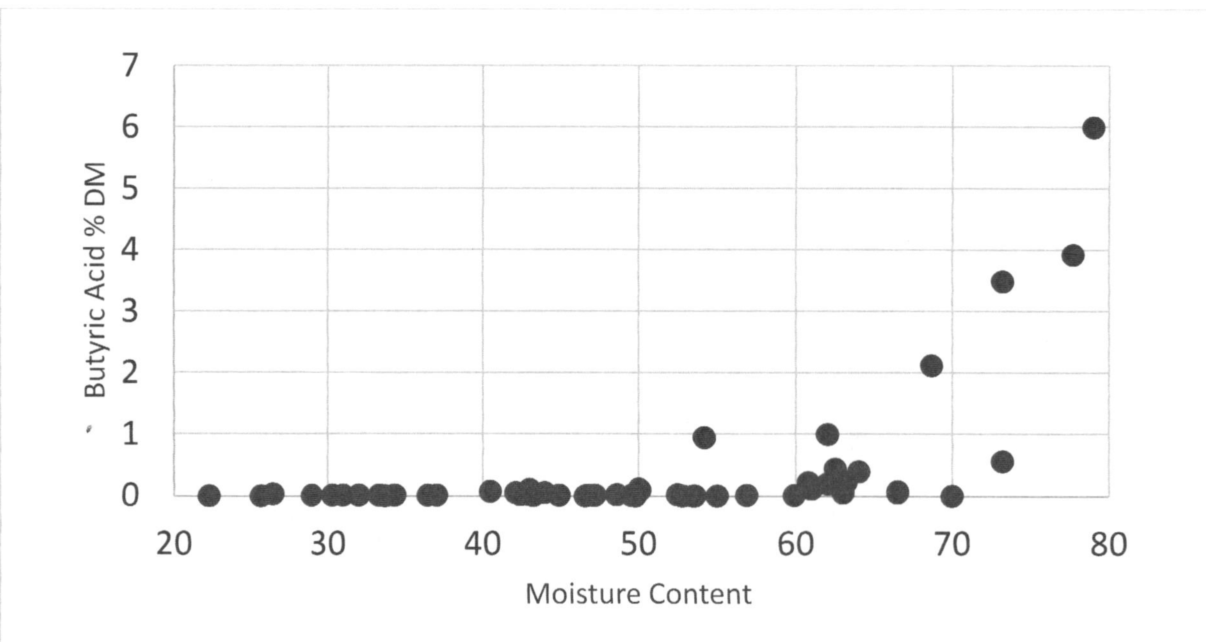

Figure 3. The effect of moisture content on butyric acid content in 2019 Kentucky baleage.

MAPPING TNC CONCENTRATIONS OVER TIME IN N-FERTILIZED AND UNFERTILIZED *ARACHIS GLABRATA* BENTH. RHIZOMES

Kelly Thomas[1]

This study examines how total nonstructural carbohydrate (TNC) concentrations change over time in *Arachis glabrata* Benth. rhizomes for three cultivars. From April 2019 to September 2020, rhizome samples were collected each month from established *Arachis glabrata* 'Ecoturf,''Florigraze' and 'FLPP 90' pastures at the UF/IFAS North Florida Research and Education Center in Quincy, Florida. 'FLPP 90' is an unreleased ornamental experimental line of *Arachis glabrata*. Half of each collection area was N-fertilized in April of each year at a rate of 50 pounds per acre and the other half remained unfertilized. *Arachis glabrata* rhizome samples were cleaned, dried and ground with a Wiley Mill grinder set at 1mm. Prepared samples were then analyzed for TNC content at the UF Forage Evaluation Support Laboratory in Gainesville, Florida. Results of the analyses show that 'FLPP 90' rhizomes had lower TNC concentrations than 'Ecoturf' and 'Florigraze' for all months and fertilization treatments in the study, ranging from 77 mg/g to 265 mg/g. The pattern of TNC concentrations over time differed for each cultivar under the N-fertilization treatment, indicating that there was a cultivar-dependent response to N-fertilization. The timing of peak TNC concentrations differed from 2019 and 2020 for all three cultivars. Generally, TNC concentrations reach peak levels during summer and are the lowest during February/March. The results suggest that genotype and abiotic factors and the interaction between the two can influence rhizome TNC concentrations of *Arachis glabrata*.

[1]UF/IFAS North Florida Research and Education Center, Quincy, FL. M.S. in Agroecology graduate student. Email: kellymthomas@ufl.edu.

MECHANICAL WEED CONTROL IN PASTURES

T.G. Wiseman, C. Gelley, C.D., Penrose, J.S. McCutcheon, M.L. Landefeld[1]

Abstract

Maintaining a quality pastures is an essential part of all forage production systems because unwanted weeds compete for nutrients, water and sunlight needed for optimum growth. Forage growth in Ohio pastures is a critical part of farm production because grazing livestock are present on more than 36 percent of Ohio farms. Weeds can lower forage crop quality and productivity if left uncontrolled. Mechanically mowing pastures is one method producers can use if they do not want to spray herbicides. The purpose of this study is to determine if weed populations in pastured forages could be changed or reduced by varying the timing of mowing throughout the late spring and summer growing period without the use of herbicides. Treatments in this study consist of: (1) Control (no mowing), (2) June only mowing, (3) July only mowing, (4) August only mowing, (5) September only mowing, (6) June and August mowing, (7) July and September mowing, and (8) mown each month; June-July-August-September. The overall weed dry matter (DM) pounds/acre (lb/ac) shows the mowing in June or June/August were significantly higher ($P<0.10$) than all other treatments except for the July only. The June treatment also yielded the most forages of 8152.0 lbs/ac on a DM basis. Mowing pastures in June, July, August and September resulted in the lowest quantity of forages of 4878.0 lbs/ac DM.

Introduction: Ohio farms have more than 1.28 million cattle and calves, more than 117,000 sheep and lambs and more than 1.42 million acres in pastureland (Turner and Morris, 2015). Topography, soil profiles and the existence of many woodland acres limit large scale agronomic row crop production. Most of the land in the Appalachian foothills of Ohio not in woodlands is used for production of cool season grasses for grazing livestock. Grazing livestock are used to harvest forages grown in pasture fields. Forages are an environmentally friendly crop that protects soils from erosion, improve soil tilth and enhance farm profitability. Weed growth in perennial forage pastures can reduce quality of the forage, quantity of useable forage and or cause livestock mortality if poisonous weeds are present and consumed. Some weeds are not palatable so livestock reject eating them. Some weeds produce chemicals that are harmful to other forage crop plants through a process called allelopathy and this can reduce growth of forage plants that are desired in the pasture. Mechanical control most often refers to mowing with a cycle bar mower or a rotary mower (brush hog). In combination with other control methods such as good fertilizer and liming practices and herbicides, mowing can be an effective tool in weed and brush management (Schnakenberg et al., 2010). The purpose of this study is to determine if weed populations in pastured forages could be changed or reduced by varying the timing of mowing throughout the late spring and summer growing period without the use of herbicides.

---

[1] Extension Educator, Ohio State University Extension; Extension Educator, Ohio State University Extension; Professor and Educator, Associate Professor and Assistant Director, Ohio State University Extension; Ohio State University Extension; retired Assistant Professor and Educator, Ohio State University Extension, respectively.

Broadleaf herbicides that control many problem weeds in Ohio pastures also reduce or eliminate legume plants that fix nitrogen from the air to feed grass forages. Adequate available nitrogen is critical for maximum grass growth in pastures. Jim Gerrish (2011) said "For an effective legume-based N supply, pastures should maintain a 30-50% legume component. Lower legume presence will lead to chronic N deficiency in the pasture and disappointing production.

Materials and Methods: This is a thirty-two (32) plot randomized complete block design with eight (8) treatments, including a control, and four (4) replications of each treatment. Each plot is fifteen feet wide by twenty feet long with an additional one foot border along each side to allow mechanical mowing equipment to be able to pass between marker posts. The site is a predominately fescue and mixed grass pasture field and the soil test (Mehlich III) results were taken in 2015 at the beginning of the multi-year experiment. Soil pH is 7.3, Phosphorus (P) 243 ppm, Potassium (K) 401 ppm, Magnesium (Mg) 456 ppm, Calcium 3037 ppm and the CEC 19.8 (meq/100g). Forage samples were taken near the beginning of June, July, August and September in 2015 and 2016. Each plot was rated for the amount of broadleaf weed pressure contained at the time of sampling. A Likert-Type scale of 0-9 was used (where 0 amounted to no visible weeds, up to 9 meaning 90% weeds). One additional rating was taken at the beginning of October before the end of the growing season. Forage samples two feet by two feet ($4ft^2$) were hand harvested from each plot and broadleaf weed species were noted. Total weight of each sample was recorded, weed(s) removed if present and weighed, and the weed-free sample weight was recorded. Any weed(s) removed from the samples were recorded by specie(s). From each of the 32 weed-free samples a sub-sample was removed to calculate dry matter. All samples were placed in the forage dryer at 46 degrees Celsius and remained there until dry. All weights were recorded in grams at that time and calculations were subsequently made from each sample to determine dry matter per acre. Each month, after harvesting samples, cow/calf pairs were used to graze the paddock, where the plots were laid out, until the desired amount of residual remained. Cattle were then removed until the next month. After each grazing, treatment plots were cut with a rotary mower according to the plan design, making one pass over the plot cutting to a height of approximately four inches above the soil surface. This was the fourth year of a five-year study. Treatments consist of: (1) Control (no mowing), (2) June only mowing, (3) July only mowing, (4) August only mowing, (5) September only mowing, (6) June and August mowing, (7) July and September mowing, and (8) mown each month; June-July-August-September.

Results: The study shows variation between plots when looking at the existing forage late in the growing season. Physical size of some weeds standout since they have not been mowed. Canada thistle and cocklebur are the two weeds most visible in the plots. However, there are a variety of other weeds such as: burdock, dandelion, horsenettle, ironweed, plantain and smartweed. While each growing season is different, rainfall amounts from the Eastern Agriculture Research Station weather station, on location, are listed in Table 1.

Discussion: Quality forages in pastures is a key to preventing weed invasion and interference and optimizing animal performance. Weeds are opportunistic in their germination and establishment during adverse weather conditions. They favor open or disturbed areas which can be caused by livestock pugging soil when conditions in pastures are wet or over grazing. Perennial, biennial and annual broadleaf weeds can have a negative effect on livestock production, so some farm managers mow pastures to reduce targeted weeds. This on farm trial is an educational experiment to help landowners determine the best time, or times, to mow pastures if trying to reduce broadleaf weed pressure and not use herbicides.

Results from this study indicated mowing in June yielded the highest tonnage of forages, but also resulted in the most weeds as well. Mowing in June in most years would prevent grasses from producing seed heads and promote tillering. It may also be too early, opening the canopy increasing weed growth. Previous data would indicate that the June/August mowing dates may be best, but it is not significantly different than other mowing frequencies except the June only cutting (Landefeld, et al., 2017). If one can reduce weed pressure by mowing at the appropriate time, the need for herbicide applications may be reduced or eliminated. Also, when mowing multiple times a year, legume plants may thrive better and become a higher percentage of the sward over a period of time. These factors could have a positive effect on forage quality in the plots for the future. This trial will continue to determine if there will be significant differences in weed populations, and differences in the yield based on the timing of the mowing.

Conclusion: Mowing pastures once per year can be an effective in maximizing forage yield. The initial weed pressure may determine the number of times a pasture may need to be mowed. Regular mowing will also greatly reduce overall forage yield, but also had the lowest quantity of weeds in this study. Regular mowing helps prevent weeds from establishing, spreading, and competing with desirable grasses and legumes (Curran and Lingenfelter, 2009). Mowing only in June and July produce the most forages, but these treatments also produced higher quantity of weeds.

Literature Cited

Curran, W. & Lingenfelter, D. (2009). Weed Management in Pasture Systems. *Penn State University.* Available at: http://extension.psu.edu/pests/weeds/control/weed-management-in-pasture-systems/extension_publication_file

Gerrish, J. (2011). Managing Pasture Fertility, American Grazing Lands Services LLC, Available at: http://ucanr.edu/sites/placernevadasmallfarms/files/102317.pdf

Schnakenberg, T., Miller, P., Hobbs, J., Scheidt, J. & Kenyon, S. (2010). Pasture Weed and Brush Control. *University of Missouri.* Available at: http://extension.missouri.edu/stone/documents/ag/articles/PastureWeedBrushAug10.pdf

Turner, C. & Morris, B. (2015). Ohio Agriculture Statistics, 2015 Annual Bulletin. Available at: https://www.nass.usda.gov/Statistics_by_State/Ohio/Publications/Annual_Statistical_Bulletin/Master14.pdf

Landefeld, M., Penrose, C., McCutcheon, J., Wiseman, T. (2017). Timing Mechanical Weed Control in Pastures. Journal of the National Association of County Agricultural Agents. Available at: https://www.nacaa.com/journal/index.php?jid=775

Table 1. Rainfall measured in inches during the significant growth months

Year	May	June	July	August	September	5 Month Total
2019	5.04	3.86	4.10	2.42	1.94	17.36

Table 2. Total weed weight

Treatments	Weed weights lbs/ac DM
Control	374.22[c]
June	551.47[a]
July	422.70[a,b,c]
August	380.89[c]
September	395.85[b,c]
June and August	526.14[a,b]
July and September	381.98
June, July, August and September	328.31[c]

[a,b,c] Different superscripts denote significant differences (P<0.10)
LSD=141.55

Table 3. Total forage weight

Treatments	Weed weights lbs/ac DM
Control	7991.0[a]
June	8152.0[a]
July	8070.0[a]
August	6582.0[a,b]
September	7496.0[a]
June and August	7551.0[a]
July and September	7157.0[a,b]
June, July, August and September	4878.0[b]

[a,b,c] Different superscripts denote significant differences (P<0.10)
LSD=2431.

META-ANALYSIS OF THE EFFECTS OF CHEMICAL AND MICROBIAL PRESERVATIVES ON HAY SPOILAGE

M. Killerby[1], R. White[2], D. C. Reyes[1], A.Y. Leon-Tinoco[1], S. Rivera[1], H. Paz[3], J. A. Jendza[4], J. J. Romero[1]

Our objective was to conduct a meta-analysis to evaluate the effects of preservatives on hay spoilage during storage. Data evaluating chemical preservatives (50 articles) and microbial inoculants (21 articles) were analyzed as separate subsets. The effect size was calculated as standardized mean differences (SMD) between treated and untreated samples. Moderators included Preservative Class (**PC**: Propionic Acid [PropA], Anhydrous Ammonia [AA], Buffered Organic Acids [BOA], Other Organic Acids [OOA], and urea), forage type (**FT**: grass, legume, or mix); moisture class [**MC**: <20% (low moisture) or >20% (high moisture)] and application rate (% w/w fresh basis; **AR**). At an AR of 0.25%, DM loss was decreased by PropA and BOA (-5.01 and -9.80%) but increased by OOA (+1.01%), relative to untreated. However, at an AR of 1.25%, PropA, BOA and OOA decreased it by -7.16, -11.4 and -8.02%, respectively. In the case of visual moldiness, we observed that applying PropA to legume, mix, and grass hay at 1.5% AR reduces it by -56.9, -27.5, and -92.4% units but when AR was increased to 3% the results were -49.9, -31.6, and -96.4, respectively. Application of microbial inoculants did not significantly affect DM loss, visual moldiness, maximum temperature, heat degree-days or *in vitro* DM digestibility. In conclusion, treatments based in organic acids were effective at preventing hay spoilage while microbial ones were not, and legume hay was less responsive to preservatives than grass hay.

[1]Animal and Veterinary Sciences, School of Food and Agriculture, Univ. of Maine, Orono, ME. [2]Animal and Poultry Sciences, Virginia Polytechnic Institute and State Univ., Blacksburg, VA. [3]Department of Animal and Dairy Sciences, Mississippi State University, MS. [4]BASF, Florham Park, New Jersey.

MIXED PERENNIAL GRASS-LEGUME SWARDS IMPACT DEEP SOIL CHARACTERISTICS

V.A. Guerra, C.L. Mackowiak, J.C.B. Dubeux, A.R. Blount, H.L. Liao, and D. Rowland[1]

Research addressing the below-ground effects of rhizoma peanut (*Arachis glabrata*) inclusion into bahiagrass (*Paspalum notatum*) systems has primarily focused on nitrogen fertility at the soil surface. A deep soil (1 meter) investigation into monoculture and mixed sward plots found treatment differences in water infiltration and root distribution with soil depth. Water infiltration in the rhizoma peanut monoculture was over 150% greater than in the bahiagrass-rhizoma peanut mix or bahiagrass monoculture plots. Rhizome mass in the 0-6-inch depth was greater in the rhizoma peanut monoculture and bahiagrass-rhizoma peanut mix than in the bahiagrass monoculture, although root mass did not differ. Root surface area in the subsoil of the bahiagrass monoculture was 2x greater than in the bahiagrass-rhizoma peanut mix and over 4x greater than in the rhizoma peanut monoculture. Root nitrogen content was greater in the mix treatment (31 lb/acre) than in the bahiagrass monoculture (15 lb/acre). Average root carbon content among treatments was 310 lb/acre. These findings suggest mixed swards may benefit from favorable belowground attributes from each species.

[1]PhD Candidate, University of Florida (UF), Quincy, FL (v.guerra@ufl.edu); Soil Specialist, UF, Quincy, FL (echo13@ufl.edu); Forage Management Specialist, Agronomy, UF, Marianna, FL (dubeux@ufl.edu); Forage Breeder, Agronomy, UF, Quincy (paspalum@ufl.edu); Soil Microbiologist, Soil and Water Sciences, UF, Quincy, FL (sunny.liao@ufl.edu); Agronomy Dept. Chair, UF, Gainesville, FL (drowland@ufl.edu).

MOB AND ROTATIONAL GRAZING: WEIGHING YOUR OPTIONS

Eric D. Billman[1], Jessica A. Williamson[2], Kathy J. Soder[3], and Danielle M. Andreen[3]

Grazing management has come to be dominated by a handful of strategies over the last 50 years. Two of these, mob and rotational grazing, present stark contrasts in terms of methodology, along with the benefits and challenges for the plants, livestock, and producers reliant on them. There is little work directly comparing the two systems, especially in the eastern US. We compared mob and rotational grazing systems over a 4-year period using a field grazing study at the Pennsylvania State University Hawbecker Research Farm. Our results indicated that mob grazing dry matter yields declined faster than rotational grazing, and that prolonged use of mob grazing eventually results in lower annual yields (6,700 lb/ac) than rotational grazing (8,900 lb/ac). Additionally, pasture species composition was heavily influenced by grazing management system, with grasses persisting better under rotational grazing (final rotational orchardgrass stand was 79%, while mob orchardgrass stand was 23%) , and alfalfa persisting better under mob grazing (final rotational alfalfa stand was 0%, while mob alfalfa stand was 26%). Finally, forage quality was impacted by management strategy, with rotationally grazed pastures having lower fiber levels (35% ADF) than mob grazed pastures (44% ADF). While neither option is universally superior, producers may use this knowledge to decide which system is best suited for the production needs.

[1]Research Agronomist, USDA-ARS, Coastal Plain Soil, Water, and Plant Research Center, Florence, SC, 29501 (eric.billman@usda.gov); [2]Assistant Professor of Forage Management, Dept. of Plant Science, Pennsylvania State University, University Park, PA, 16802 (jaw67@psu.edu); [3]Research Animal Scientists, USDA-ARS, Pasture Systems and Water Management Research Unit, University Park, PA. 16802 (kathy.soder@usda.gov; elle.andreen@usda.gov).

NOVEL PEST MANAGEMENT STRATEGIES TO CONTROL SMUTGRASS IN SOUTHERN FORAGE SYSTEMS

Nicholas J. Shay, Lisa L. Baxter, Nicholas T. Basinger, and Brian M. Schwartz[1]

Smutgrass (*Sporobolus indicus*) can quickly outcompete bahiagrass (*Paspalum notatum*) because of its aggressive growth, prolific seed production, and rhizomatous growth. Total renovation of a bahiagrass pasture or hayfield is generally not a feasible or economically viable option for most producers. Therefore, controlling the continual spread of smutgrass will require an integrated pest management plan that incorporates multiple strategies. The objective of this study is to test the interactions of multiple IPM strategies including herbicides and fertilizers on smutgrass control in bahiagrass. This research was conducted on mixture of Tifton 9 and Pensacola bahiagrass at the Alapaha Beef Station in Alapaha, GA. Plots were arranged in a three by four factorial design with six replications. Herbicide treatments included Indaziflam (Rezilon, PRE), Hexazinone (Velpar, POST), combination of pre + post-emergent herbicides (PRE + POST), and an unsprayed control. Fertilizer treatments included 50 units N/acre (ammonium nitrate, 34% N) + 50 lb K_2O/acre, 50 units N/acre (N), and an unfertilized control. Preliminary results show that plots receiving PRE and POST emergent herbicides in addition to N and K_2O resulted in an improved bahiagrass stand as timely weed suppression removed competition, while fertilizer provided essential nutrients for optimum growth to fill in the gaps. This work will be repeated in 2021 before final conclusions are made.

[1]Graduate Student, University of Georgia, Crop and Soils Science Dept., University of Georgia, Tifton, GA (nicholas.shay@uga.edu); State Forage Extension Specialist, Crop and Soils Science Dept., University of Georgia, Tifton, GA (baxterl@uga.edu); Assistant Professor, Crop and Soils Science Dept., University of Georgia, Athens, GA (nicholas.basinger@uga.edu); Professor, Crop and Soils Science Dept., University of Georgia, Tifton, GA (tifturf@uga.edu).

ON-FARM DEMONSTRATIONS MISTAKES MADE & LESSONS LEARNED

M.D. Mauldin[1]

On-farm demonstrations are an excellent means by which Extension faculty can answer complicated questions for cooperating clients, effectively educate the public, and gain valuable field experience. Over the past five years, the Agent has conducted 15 on-farm demonstrations (six pasture herbicide and 9 forage variety demonstrations) in Washington County, FL. Some demonstrations have been quite successful – meaningful for the agent, cooperating clients, and the public, while others can only be classified as failures – generating no meaningful information for anyone involved. Experience and reflection have revealed four key factors necessary for successful on-farm demonstrations. **1) Communication** – Communication between Extension faculty and cooperating clients must begin before the project starts, be maintained throughout the entire process, and be a two-way street. **2) Follow Through** – Demonstrations involving forage systems generally take quite some time to complete. Both parties must be willing to complete the entire project for the experience to have real meaning. **3) Site Selection** – If a demonstration is not implemented on an appropriate site it has no chance to be effective. Specific characteristics of an appropriate site will vary, but factors including (not limited to) soil productivity, ability to exclude grazing animals, site history (and future plans), and accessibility should be carefully evaluated before beginning the demonstration process. **4) Technical Issues** – While the technical aspects of a demonstration must be handled appropriately, these issues only have bearing on the overall success of the demonstration if the first three factors are properly addressed. Technical competence alone is not sufficient to ensure the success of the demonstration.

[1]Agriculture & Natural Resources Agent, University of Florida IFAS Extension, Washington County, Chipley, FL (mdm83@ufl.edu).

OUR COMPREHENSIVE PASTURE HERBICIDE STEWARDSHIP PROGRAM AT THE UNIVERSITY OF TENNESSEE

G.N. Rhodes, Jr., L.S. Steckel, T.C. Mueller, and D.W. McIntosh[1]

Due to the sensitivity of many high value crops and other plants to off-target auxinic herbicides, we began a comprehensive educational program in 2011 that stresses the importance of proper stewardship with the use of pasture herbicides. Our goals were to reduce the occurrence and impact of off-target damage to tobacco and other sensitive, high value crops; and to create educational materials and other tools to help with the diagnosis of suspected cases of off-target damage. We focused on four crops (tobacco, cotton, tomato and grape) and five herbicides (2,4-D, dicamba, aminopyralid, aminocyclopyrachlor and picloram) for the creation of educational materials and diagnostic tools. These include still images, time lapse videos and fact sheets, and we made them available through our website, herbicidestewardship.com it became accessible in 2014. In 2016 and again in 2017, severe problems with dicamba drift occurred in the Midsouth on numerous sensitive crops as a result of in-crop applications of the herbicide in dicamba-tolerant cotton and soybean varieties. In 2017 we broadened our website to include al information addressing stewardship of dicamba and 2,4-D tolerant crop technology.

[1]Professor, The University of Tennessee, Knoxville, TN (nrhodes@utk.edu); Professor, The University of Tennessee, Knoxville, TN (tcmueller@utk.edu); Professor, The University of Tennessee, Jackson, TN (lsteckel@utk.edu); Research Coordinator, The University of Tennessee, Knoxville, TN (dmcintos@utk.edu).

PASTURE MANAGEMENT CURRICULUM FOR EXTENSION AGENTS

C. Bainum and M. Wallau[1]

This project was created and developed to support continuous improvement in pasture management subject matter programming to enhance agent training and producer support. Within the organization of UF/IFAS there are county faculty charged with agronomic responsibilities that extend beyond their level of knowledge and/or training. This results in reluctance to deliver programs in those areas where training is limited due to lack of agent confidence in the subject matter. The need for pasture management classes on a more regular basis is evidenced by annual evaluations and post-program surveys that reveal pasture management topics of greatest importance to producers. The objective of this project was to streamline pasture management topics into a state-wide curriculum with credible supplemental resources for each major topic to provide UF/IFAS Extension agents the framework to develop their own teaching tools to better serve Florida livestock producers. The pasture school curriculum is organized into four learning modules. Thinking with the end goal in mind, the selling of a final livestock product i.e. a load of calves from the Southeast, the modules are focused on forage establishment and production, management, and longevity specific to forages of Florida and how to maximize utilization during each phase of livestock production. This curriculum was newly developed in 2020 and was used for regional programming in 2020 inflicting practice change in 34 producers.

[1] Livestock Extension Agent, UF/IFAS, Marion County Extension, Ocala, FL (cbainum@ufl.edu) & Forage Extension Specialist, UF/IFAS, Gainesville, FL (mwallau@ufl.edu).

PEANUTS: A NEW FORAGE FOR THE SOUTHERN GREAT PLAINS?

T.W. Witt and N. Puppala[1]

Peanuts (*Arachis hypogaea* L.) are grown throughout the southern U.S. and the world as a food and oil crop. Throughout the Southern Great Plains of the U.S. there is a growing interest to examine peanuts as a source of forage. In this study, ten peanut genotypes representing four market classes (Runner, Spanish, Valencia, and Virginia) were grown at two locations (Clovis, NM and El Reno, OK) to determine the biomass production and forage quality [acid detergent fiber (ADF), crude protein (CP), in vitro true digestibility (IVTD), neutral detergent fiber (NDF), percent carbon (C), percent nitrogen (N), and relative feed value (RFV)] for late season peanuts (>15 weeks after planting). In 2019 and 2020 peanuts were harvested at multiple time points after 15 weeks of growth until a killing frost. After harvesting the peanuts, a fresh weight and dry weight were taken then the samples were ground to 0.08 inches using a Wiley mill for laboratory analysis. ADF and NDF were evaluated using the ANKOM method for a batch fiber analyzer. Total N and C were determined using an elemental analyzer. CP was calculated by multiplying the %N by 6.25. IVTD was obtained by using the procedures for a Daisy incubator. The dry matter production ranged from 4879 lb/acre to 4094 lb/acre with RFV scores ranging from 127 to 120 and crude protein ranging from 1.1 oz/lb to 0.9 oz/lb. Analyses of the data revealed that the four market types had statistically similar ($P \leq 0.05$) biomass production. However, the market types were significantly ($P \leq 0.05$) different for the quality traits of CP, NDF, and RFV.

[1]Research Agronomist, USDA-ARS, Grazinglands Research Laboratory, El Reno, OK (travis.witt@usda.gov); Peanut Breeder, NMSU, Agricultural Science Center at Clovis, Clovis, NM (npuppala@ad.nmsu.edu).

PERCEPTIONS OF ALFALFA ADOPTION IN THE SOUTHEAST US

L.S. Silva, M.K. Mullenix, J.J. Tucker, and C.G. Prevatt[1]

Alfalfa was once a dominant forage in the Southeast US, but the harsh environment and insect pressure promoted its acreage decline. Recent breeding efforts developed new varieties that are more tolerant to growing conditions in the region and can be used under grazing management. A recent network collaboration led by University of Georgia with Auburn University and University of Florida is leading efforts to incorporate alfalfa back into forage systems. In an attempt to better understand main challenges and limitations of alfalfa adoption in the Southeast US, a survey was released in May 2020 targeting hay and beef producers that were interested in or already planting alfalfa in Alabama, Arkansas, Florida, Georgia, Kentucky, Lousiana, Mississipi, North Carolina, South Carolina, Tennessee, Virginia or Texas. The results (n= 207 respondents) showed approximately 85% participants have considered growing alfalfa and 65% participants already invested time in learning more about alfalfa, including accessed Extension publications. However, only 21% of total participants (n= 40) are currently growing alfalfa (up to 75 acres) and 61% in monoculture. Within this group, they reported that alfalfa adoption in their forage systems increased forage yield, hay quality and profit, but also pest and weed occurrence. This was also one reason why some producers stopped growing alfalfa, along with weather challenges and higher management effort or cost. In conclusion, alfalfa adoption can help increase forage yield and quality and decrease off farm inputs on forage-based systems, but it requires closer management to improve its longevity and persistence on pastures.

[1]Post-Doctoral Research Fellow, Auburn University, Auburn, AL (lzs0105@auburn.edu); Associate Professor, Auburn University, Auburn, AL (mullemk@auburn.edu); Assistant Professor, University of Georgia, Tifton, GA (jjtucker@uga.edu); State Specialized Extension Agent II, University of Florida, Ona, FL (prevacg@ufl.edu).

PLANT GROWTH-PROMOTING RHIZOBACTERIA (PGPR): EFFECTS ON 'KY 31' TALL FESCUE AND 'RUSSELL' BERMUDAGRASS NUTRITIVE VALUE AND YIELD

M.L. Cole[1], S.L. Dillard[1], R.B. Muntifering[1], M.K. Mullenix[1], and D.W. Held[2]

Additional N sources ensure economic viability of forage systems due to high N cost. This trial evaluated plant growth-promoting rhizobacteria (PGPR) as possible N sources for 'Russell' bermudagrass (*C. dactylon*) and 'KY 31' tall fescue (*L. arundinaceum*). Fourteen, 32.3-ft^2 plots treated with full N (50 lb N/acre), ½ N (½ N; 25 lb N/acre), Accomplish LM (AMS), AMS + ½ N, DH44, Blend 20 and control. Samples were taken every 4 weeks with 1.1-ft^2 quadrats and analyzed for NDF, ADF, CP, and DM using NIRS; 10% were randomly selected for wet chemistry. Data were analyzed using Proc GLIMMIX of SAS 9.4 (SAS Inst., Cary, NC), a completely randomized design. Full N had greatest yield (7,401 lb DM/acre) for 'KY 31' compared to control (6,560 lb DM/acre; $P < 0.01$) and DH44 (6,591 lb DM/acre; $P < 0.02$). Full N had greatest yield (8,323 lb DM/acre; $P < 0.02$) compared to control (7,325 lb DM/acre) for 'Russell'. No differences reported ($P \geq 0.05$) in NDF for both forages. Full and ½ N had greater (11.5, 11.4%; $P \leq 0.03$) CP than AMS in 'Russell'. Blend 20 treated 'Russell' had higher (32.6%) ADF than full N (30.7%; $P < 0.02$) and ½ N (30.9%; $P < 0.03$). Full N had lower (30.7%; $P < 0.04$) ADF than AMS + ½ N (32.4%) for 'Russell'. Blend 20 treated 'KY 31' had greater (35.8%) ADF compared to AMS + ½ N (34.5%; $P < 0.02$), DH44 (33.9%; $P < 0.01$), full N (34.3%; $P < 0.01$), and control (34.2%; $P < 0.01$). Full N had greater CP (12%; $P < 0.01$) compared to AMS in 'KY 31'. DH44 treated 'KY 31' had greater CP compared to AMS and Blend 20 (11.9% vs. 11.0%; $P < 0.04$). PGPR treated plots produced yield and nutritive quality similar to commercial N source.

[1]Department of Animal Sciences, Auburn University, Auburn, Alabama .

[2]Department of Plant Pathology and Entomology, Auburn University, Auburn, Alabama

POTASSIUM FERTILIZATION AND ITS IMPACT ON YIELD, NUTRITIVE VALUE, AND WINTER HARDINESS OF ALFALFA

A. Greenberg, M.T. Berti, D. Franzen, A. Peterson, A. Wittenberg, S. Cabello, K. Mozea, and S. Bibby[1]

Medicago sativa L., commonly called alfalfa, is the most widely grown forage crop in the world. An important nutrient for improving alfalfa's yield, quality, and winter hardiness is potassium. There are multiple factors that should determine the appropriate rate of potassium fertilization, one being the soil's clay mineralogy. When the soil's smectite-to-illite clay ratio is greater than 3.5, potassium becomes immobile within the soil and is less accessible to the crop. The experiment was conducted at Milnor and Lisbon, ND, in 2019, both fields had low K in the soil (<100 mg/kg K) and two different smectite-to-illite ratio (high and low). Three alfalfa varieties with fall dormancy scores of 3, 4, and 5 were fertilized with five different treatments of potassium. The treatments were 0, 168, and 336 kg K_2O/ha at seeding, and 168 and 336 kg K_2O/ha as split-application. There were no significant effects of K fertilization on total forage yield in the seeding year ($P = 0.05$). The FD3 variety had a lower yield, which was expected. For nutritive value, ash content was significantly higher and TDN was significantly lower with higher rates of K. This study formed a stronger understanding of the most appropriate potassium rates for alfalfa to maximize quality, yield, and winter survival. By better understanding the most efficient rates of potassium application, depending on the soil's clay mineralogy, growers can more economically, produce higher quality and yield of alfalfa.

[1]Graduate Research Assistant, MS student (amy.greenberg@ndsu.edu); Forage and Biomass Production, Professor (marisol.berti@ndsu.edu); Soil Science, Professor (david.franzen@ndsu.edu), Research Specialist (alan.peterson.2@ndsu.edu); Research Specialist (alex.1.wittenberg@ndsu.edu), PhD student (sergiofabian.cabello@ndsu.edu); MS student (kenneth.mozea@ndsu.edu); MS student (samuel.bibby@ndsu.edu) North Dakota State University, Fargo, ND.

POTENTIAL USE OF BERMUDAGRASS TURFGRASS IN SOUTHERN AGRICULTURE SYSTEMS

Lisa Baxter, Leanne Dillard, Katie Mason, and Brian Schwartz[1]

Regardless of size, nearly all livestock or horse farms in the Southeastern US have high-traffic areas that are prone to erosion such as areas surrounding hay rings, water troughs, or working facilities. Unfortunately, traditional forage grasses cannot withstand continual hoof or tire wear and will begin to thin, even under optimal weather conditions. Three forage-type bermudagrass (*Cynodon dactylon*) varieties were compared to a turf-type bermudagrass under intensive simulated grazing and hay management in small plot trials established in Tifton, GA and Auburn, AL. Varieties included: Alicia, Coastal, Coastcross II, and DT-1 (turf-type). Grazed treatments were clipped to ½ in. stubble and hay treatments to a 2 in. stubble height. Coastcross II produced the greatest yields under intense grazing management, while the other varieties produced similar yields ($P < 0.01$). The varieties responded differently to hay management in each location ($P < 0.01$). Only Coastcross II yields were greater than DT-1 in Auburn ($P = 0.03$), whereas the other varieties were intermediate and similar to both ($P > 0.05$). Coastcross II generated the greatest yield in Tifton followed by Coastal, Alicia, and DT-1 ($P < 0.01$). DT-1 visually maintained greater ground coverage, which will be corroborated by digital image analyses this fall. The integration of turfgrass appears to be a viable option for the humid south under intensive grazing conditions and high-traffic areas of livestock or equine production systems.

Forage Extension Specialist, Crop and Soils Science Dept., University of Georgia, Tifton, GA (baxterl@uga.edu); Forage Extension Specialist, Animal Science Dept., Auburn University, Auburn, AL (dillasa@auburn.edu); Beef Extension Specialist, Animal Science Dept., University of Tennessee, Knoxville, TN (kmason21@tennessee.edu); Professor, Crop and Soil Science Dept., University of Georgia, Tifton, GA (tifturf@uga.edu).

SERICEA LESPEDEZA—THE RODNEY DANGERFIELD OF THE FORAGE WORLD

Reed Edwards

Sericea Lespedeza is a plant with a long, somewhat interrupted use as a forage plant. But, due to rising fertilizer costs in recent years, and the increase in numbers of small ruminants, it has risen dramatically in use in recent years. Still, it is rarely mentioned in the company with other premium forage plants. Sericea is very drought tolerant, tolerates low fertility, and low soil pH. It is very effective in the battle against *Haemonchus contortus* and coccidia in small ruminants and camelids. It is insect resistant, makes very high quality hay with a bit of care, and due to its quick drying characteristics, allows a farmer a good chance of putting up high quality hay even when the weather in unfavorable to make other types of hay. With grazing-tolerant varieties, it is time that Sericea Lespedeza got a little more notice in the forage world.

SHEEP AND GOAT PREFERENCES OF FIVE COMMON COVER CROPS

Robert A. Lane, Jaclyn Robinson, and Shyam Nair[1]

While the use of cover crops continues to gain widespread acceptance, producers frequently seek other means of deriving economic value from such crops, such as their use cover as livestock forage. We evaluated sheep and goat intake and apparent preference when provided free access to freshly harvested cereal rye (*Secale cereale* cv. "Elbon"), annual ryegrass (*Lolium multiflorum*), berseem clover (*Trifolium alexandrinum* cv. "Frosty"), daikon radish (*Raphanus sativus* subsp. *longipinnatus*), and Austrian winterpea (*Pisum sativum*) with advancing plant maturation (March 5 to 27, 2020). While there was significant interaction between sheep and goats for dry matter (DM) consumption over time, both species preferred ryegrass over daikon radish. However, total DM consumption of ryegrass was not greater than that of the other crops. Goats spent significantly more time at ryegrass, rye, and berseem clover treatments (troughs filled with fresh forage) while sheep spent more time on berseem clover and ryegrass. Austrian winterpea and daikon radish were the least preferred forages. Animals spent significantly more time on ryegrass at the early feeding trial and significantly more on berseem clover at the last feeding trial 22 days later. Legumes had higher crude protein and lower ADF values than non-legume species throughout the study, but there was no significant correlation between forage quality and animal preference. While other factors may that limit use of these species as cover crops, our results indicate that ryegrass, berseem clover and rye have good potential to serve as forage for small ruminants while daikon radish and Austrian winterpea are less suitable.

[1]Professor, School of Agricultural Sciences, Sam Houston State University, Huntsville, TX (blane@shsu.edu); Graduate Assistant, Department of Agricultural Economics, Texas A&M University, College Station, TX (jaclynrobinson@email.tamu.edu); Associate Professor, School of Agricultural Sciences, Sam Houston State University (shyam.nair@shsu.edu).

STEM-TO-LEAF RATIO: RECALCULATING CONCEPTS OF FORAGE QUALITY

Gonzalo Ferreira, Christy L. Teets, and Matias J. Aguerre[1]

There are two major determinants of the nutritional quality of forages. The first determinant of forage quality is the concentration of neutral detergent fiber in plant tissues. The second determinant of forage quality can be the leaf-to-stem ratio within the forage. In general terms, forages with greater leaf-to-stem rations contain greater coefficients of dry matter (DM) and NDF digestibility given that leaves have lower concentrations of NDF. However, is the latter concept valid for (most) summer annual grasses? In a recent study, we harvested leaves and stems from summer annual grasses (corn, sorghum, and pearl millet), alfalfa, and winter annual grasses (barley, oats, rye, ryegrass, triticale, and wheat) and determined their NDF concentration. The concentrations of NDF (DM basis) in leaves and stems of alfalfa were 26.3 and 58.7%, respectively, and the concentrations of NDF in leaf blades and stem internodes of winter annual grasses were 44.7 and 60.7%, respectively. Leaf tissues had lower concentrations of NDF than stem tissues. However, the concentrations of NDF in leaf blades and stem internodes of summer annual grasses were 56.0 and 45.8%, respectively. These data imply that a greater leaf-to-stem ratio might not necessarily result in lower NDF concentrations for the whole plant. The lower concentration of NDF in stem internodes relative to leaf blades of summer annuals has been observed in several other studies from our research group. For certain summer annual grasses, a greater leaf-to-stem ratio may not result in forages of less NDF concentration.

[1]Dairy Management Extension Specialist, Department of Dairy Science, Virginia Tech, Blacksburg, VA (gonf@vt.edu); Laboratory Research Assistant, Department of Dairy Science, Virginia Tech, Blacksburg, VA (cteets77@vt.edu); Ruminant Nutritionist, Animal and Veterinary Sciences Department, Clemson University (maguerr@clemson.edu).

THE EFFECT OF IRRIGATION RATE ON THE YIELD AND SPECIES COMPOSITION OF TWO PERENNIAL GRASS/LEGUME MIXTURES IN SOUTHERN OREGON

G.B. Jones[1]

Management decisions play an important role in the yield and composition of perennial forage mixtures. While the yield response of forage crops to drought is understood, the effect of water stress on the species composition of forage mixtures has not been well studied. The aim of this study was to determine the yield and species composition of two perennial forage mixtures irrigated with a line-source sprinkler design in the Mediterranean climate of southern Oregon. Two mixtures (1) tall fescue and birdsfoot trefoil, and (2) tall fescue, orchardgrass, birdsfoot trefoil, white clover, red clover, and plantain were planted in replicated plots in the fall of 2019 (n = 4). In May of 2020, a line-source sprinkler system was installed which provided a linearly-decreasing application of irrigation water to the plots. In July and September, two 0.1 m^2 quadrats were harvested from plots which had been irrigated to replace 100%, 66%, 33% and 0% of calculated evapotranspiration. Yield of both mixtures at both cuttings declined linearly with decreasing irrigation application ($P < 0.01$). At the July harvest, the relative composition of mixture 1 did not change with irrigation application, but in mixture 2, the proportion of red clover ($P = 0.002$) and plantain ($P = 0.03$) biomass increased with increasing irrigation while the proportion of non-planted weeds decreased ($P = 0.003$). The relative proportions of other species in mixture 2 did not change with irrigation application rate. Further data will be collected through the 2021 growing season to evaluate competition among and persistence of species as affected by irrigation rates.

[1]Assistant Professor (Practice), Oregon State University, Southern Oregon Research & Extension Center, Central Point, OR (gordon.jones@oregonstate.edu).

THE EFFECTS OF COVER CROPS IN AN INTEGRATED LIVESTOCK/CONTINUOUS CORN CROPPING SYSTEM IN EAST-CENTRAL MISSISSIPPI

K.R. Waddell[1], J.B. Rushing[2], B.S. Bass[3], and J.C. Lyles[4]

There is a growing interest within the agricultural industry on how cover crop systems affect soil health and economic productivity in integrated crop-livestock systems. This study investigates the combined effects of cover crop, tillage, and grazing in a continuous corn (*Zea mays*) production system in east-central Mississippi. This study consists of a randomized complete block design, with two tillage treatments (conventional and no-till), twelve cover crop treatments, and separate grazed and un-grazed plot areas. Cover crop treatments include mono- and polycultures of oats (*Avena sativa*), cereal rye (*Secale cereale*), crimson clover (*Trifolium incarnatum*), and radish (*Raphanus sativus*), along with a control. In the grazed area, bred heifers are used to graze cover crop treatments in a cafeteria style design. Data collection includes forage mass (FM), nutritive value (crude protein – CP; total digestible nutrients – TDN), and soil analysis (nutrient availability, compaction, bulk density, and moisture). Crude protein ranged between 11.5% to 14.9% across both trials. Total digestible nutrients were significantly greater for the first sampling date compared to the final three sampling dates. The un-grazed area had significantly greater FM compared to the grazed area for all four sampling dates ($P<0.0001$). Mean organic matter was significantly greater for the grazed area (1.11%) compared to the ungrazed area (0.87%) ($P<0.0001$). There were no differences between grazed and un-grazed areas for mean soil compaction across all sampling depths ($P=0.0718$).

[1]Graduate Assistant, MSU-CPBES, Newton, MS (krw324@msstate.edu); [2]Assistant Research/Extension Professor, MSU-CPBES, Newton, MS (brett.rushing@msstate.edu); [3]Graduate Assistant, MSU-CPBES, Newton, MS (bsb341@msstate.edu); [4]Research Associate I, MSU-CPBES, Newton, MS (jcl411@msstate.edu).

THE EFFECTS OF GRAZING COVER CROPS ON ANIMAL PERFORMANCE, FORAGE MASS, AND NUTRITIVE VALUE IN EAST CENTRAL MISSISSIPPI

B.S. Bass, J.B. Rushing, K.R. Waddell, and J.C. Lyles

Integrated crop-livestock systems have been widely used throughout history, allowing for multiple agricultural practices on a single tract of land. With a growing awareness in recent years regarding soil health and agricultural sustainability, these systems have begun to be re-introduced into the southeastern U.S. Research for this project began in the fall of 2019 and will be continued over a 2-year period at the Coastal Plain Branch Experiment Station (CPBES) in Newton, MS evaluating animal performance, forage mass, and nutritive value while grazing cover crops. The following three cover crop treatments were planted in nine, two-acre paddocks in a randomized complete block design in Oct of 2019: (O) oats (*Avena sativa*), (OC) oats + crimson clover (*Trifolium incarnatum*), and (OCR) oats + crimson clover + daikon radish (*Raphanus sativus*). Commercial angus (*Bos taurus*) steers (mean starting weight 510 lb) were stocked at 1020 lb/a for a 66-d grazing season. Steers were weighed every 28-d. Average daily gain (ADG) was significantly less (P=0.0108) in the OCR treatment (2.69 lb/hd/d) when compared to O (3.67 lb/hd/d) and OC (3.25 lb/hd/d). Total gain per acre (GAIN) was significantly greater (P=0.1063) in O (619 lb/a) than in OCR (478 lb/a). Forage mass (FM) for samples collected bi-weekly during the grazing season were greatest for O (2056 lb/a), followed by OC (1851 lb/a), and OCR (1774 lb/a; P=0.0297). Crude protein (CP) in OCR (17.8%) was significantly greater (P=0.0084) than O (16.3%). Oats had significantly greater mean total digestible nutrients (TDN; 61%) compared to the other two treatments (P<0.001), across all sampling dates.

[1]Graduate Assistant, MSU-CPBES, Newton, MS (bsb341@msstate.edu); [2]Assistant Research/Extension Professor, MSU-CPBES, Newton, MS (brett.rushing@msstate.edu); [3]Graduate Assistant, MSU-CPBES, Newton, MS (krw324@msstate.edu); [4]Research Associate I, MSU-CPBES, Newton, MS (jcl411@msstate.edu)

THE HISTORY OF THE INTERNATIONAL GRASSLAND CONGRESS 1927-2020

S. Ray Smith, Vivien Allen, Roger Wilkins, and Garry Lacefield[1]

Grasslands cover 27% of the world's land area and 70% of agricultural land. They form a foundation for agriculture and livestock husbandry and contribute to the livelihoods of more than 800 million people. The International Grassland Congress (IGC) began as a meeting of a small group of forage researchers in Europe in 1927. It expanded into a gathering held every 3-5 years with over 1000 delegates from over 80 countries representing research, extension, and production. The newly-released IGC history provides not only an overview of the IGC Congress over the last 93 years, but highlights the people, events, and changes in grassland agriculture around the world. The current edition is an expansion and update of 'A Brief History of the International Grassland Congress' presented at the XX IGC in Ireland in 2005 by Professor Ross Humphreys, Australia. The authors worked diligently to highlight the individuals that shaped the IGC since its inception, including excerpts of presentations by world grassland leaders. They even included part of the welcome address from Prince Philip, Duke of Edinburgh, who was President of the XIII IGC held in Reading, United Kingdom, in 1960. The IGC History describes the two Congresses held in the United States, the XI IGC in State College, PA in 1952 and the XIV IGC in Lexington, KY in 1981. It also provides details on the formation of the Forage and Grassland Foundation by Dr. John Baylor and Mr. Warren Thompson from the $100,000 profit of the XIV Congress. The IGC History will be available for viewing and downloading on the IGC website early in 2021 at https://www.internationalgrasslands.org/publications.

[1]Professor and Forage Extension Specialist, University of Kentucky, Lexington, KY (raysmith1@uky.edu); Paul Whitfield Horn Professor Emeritus, Texas Tech University; retired Deputy Director of the Institute of Grassland and Environmental Research, United Kingdom; and Professor Emeritus, University of Kentucky, Bowling Green, KY.

THE USDA GRAZE PASTURE BUDGETING MODEL FOR KENTUCKY: A NEW, REVISABLE EDITION

Jimmy C. Henning, Ivelin Denev, S. Ray Smith, Chris Teutsch, and Adam Jones[1]

The ability to estimate stocking rate as well as monthly and yearly forage balances is helpful to forage livestock producers. The Kentucky USDA GRAZE pasture budgeting model is a simple, spreadsheet-based tool that allows input of forage, livestock, soil, management and utilization parameters to estimate monthly and annual forage balance. The model is based on 'look up' tables for forage productivity by month, yield, and adjustments according to soil suitability group. Forage intake is a simple calculation using static body weight of livestock and intake as a percent of the body weight. The tool is designed to be a farm planning tool to show the benefit from changing the forage base, management, and percentage utilization (from rotational grazing). The model allows for multiple livestock types, including cattle, horses, sheep and goats. In 2020, the Kentucky GRAZE model was updated to improve the user interface and to incorporate new forage species and yield data into the database. The revised GRAZE model is a Microsoft Excel spreadsheet that will run on current and older Windows-type personal computers as well as Apple iPads. The forage species and yield data were compiled from University of Kentucky trials and other sources and results are applicable to the transition zone of the Southeastern United States. The model is designed to allow future incorporation of updated forage and yield data as needed.

[1]Extension Professors (jimmy.henning@uky.edu and raysmith1@uky.edu), Associate Extension Professor (chris.teutsch@uky.edu), University of Kentucky, Lexington, KY 40546 and USDA NRCS Natural Resources Specialist (adam.jones2@usda.gov) USDA NRCS, Lexington, KY 40503.

TRENDS AND BARRIERS FOR HIGH-QUALITY FORAGE VARIETIES

J.H. Bouton[1]

Persistent grass species form our main forage base, but most are moderate to low in nutritive quality; even toxic (tall fescue). Public and private forage breeders over time 1) developed new varieties of the low-quality grasses with incrementally better quality and 2) added persistence to high quality grasses that allowed more of their general use. Other programs developed persistent legume varieties to mitigate grass nutritive quality problems when inter-planted. Traditional selection and breeding techniques were used for these advancements although a biotech approach was employed recently to develop reduced lignin alfalfa. Finally, elimination of toxic tall fescue pastures and replacement with newly developed non-toxic or "novel" endophyte varieties is the best opportunity to solve "fescue toxicosis" and increase producer profit. Despite many educational programs for over 20 years, and based on current seed sales, farmers are not planting novel endophyte varieties in appreciable quantities. Along with normal risk-averse management inherent among forage-livestock producers for making changes with costly new products, ownership and confirmation biases are additional barriers that need to be overcome: "I own my current fescue stand, it is green and growing, and I have used it successfully, so it is valuable and in no need of being replaced or even mitigated". Intensive educational programs and on-farm demonstrations that disrupt these biases and show better animal health and profit are now being employed by industry and extension partnerships to increase use of novel endophyte varieties and other high-quality forage varieties.

[1] Professor Emeritus, University of Georgia, and Owner, Bouton Consulting Group, LLC, Athens, GA 30606 (jhbouton1@gmail.com).

UK MASTER GRAZER: FACILITATING CHANGE THROUGH ON-FARM DEMONSTRATIONS

C.A. Thrailkill and C.D. Teutsch[1]

UK Master Grazer is an outreach program at the University of Kentucky that is funded through a grant from the Kentucky Agricultural Development Fund. This program provides producers with workshops and on-farm demonstrations designed to facilitate the adoption of improved grazing management. Since farmers tend to be experiential learners, on-farm demonstrations are a key activity of the Master Grazer program.

- *Crabgrass.* Demonstrations of improved crabgrass varieties were conducted on two farms in Western Kentucky. Producers were able to see and touch crabgrass being utilized as forage.
- *Summer annual mixtures.* Demonstrations included monocultures, and simple and complex mixtures of summer annual grasses, legumes and fobs on seven farms in Western KY.
- *Reclamation of Hay feeding areas.* Areas where hay was fed the previous winter are often infested with weeds during the summer months. This demonstration showed how these areas could be revegetated using a diverse summer annual mixture.
- *Getting started with rotational stocking.* A simple rotational stocking system was established on two farms in Western KY using temporary fencing and a solar charger.

These demonstrations have facilitated the adoption of improved forage and grazing management on farms in Kentucky and surrounding states.

[1]Master Grazer Coordinator, University of Kentucky Research and Education Center, Princeton, KY 42445; and Extension Associate Professor, University of Kentucky Grain and Forage Center of Excellence, Princeton, KY 424454.

WATER QUALITY FOR GEORGIA LIVESTOCK

C.H. Knight[1], M.A. Wunderly[2], R.G. Fitzpatrick[3], L.E. Ray[3], T.W. Cheely[3], G.P. Pittman[3], T. Glenn[3], S.R. Patrick[3], S. Reynolds[3], C.M. Stephens[3], R.L. Stewart[3], and J. Burnsed[3]

Livestock production relies on good water quality sources for health, reproduction, and adequate consumption. Water sampling was conducted in northeast Georgia to test parameters impacting animal health (Fe, MN, sulfate, Mo, Cr, Ca, As, Se, Cd, Pb, Cu, Na, fluoride, chloride, phosphate, soluble salts, total dissolved solids, turbidity, nitrate, nitrite, pH, fecal coliform bacteria, and water temperature). Data loggers recorded water temperature every 15 minutes in 30 livestock watering sources during August, September and October. Twenty-five farms participated and 50 watering sources were sampled, including well water troughs, ponds, and streams. Chemical elements were within recommended limits for 60% of samples. The most common elemental contaminants above the limit were iron (22%), manganese (18%), and pH (12%). Two samples had high nitrate-nitrogen levels, and copper and phosphorous were each higher than recommended in one sample. Turbidity levels were unacceptable in one stream and one pond. Fecal coliform levels were higher than recommended in 86% of the samples, with highest counts from troughs in high density corrals. Water temperatures exceeded the 80° F recommended limit 0% of the time for streams, 17% for troughs, and 52% for ponds. Temperature, fecal coliform counts, and pond/stream turbidity can change more frequently than elemental contaminants, requiring additional monitoring. Fecal coliform contamination is the greatest concern for livestock water quality in northeast Georgia, followed by pond temperature.

[1] County Extension Coordinator and Agriculture and Natural Resource Agent, University of Georgia Cooperative Extension, Madison County, Danielsville, GA, clh@uga.edu, [2]Area Water Agent, University of Georgia Cooperative Extension, Northeast District, Watkinsville, GA, martin.wunderly@uga.edu; [3]Agriculture and Natural Resource Agents, UGA Cooperative Extension, Northeast District.

WET FEET, DRY FEET, HOT FEET:
WHICH FORAGES MIGHT TOLERATE ALL THREE?

S.R. Freeman and M.H. Poore[1]

Hurricanes frequently bring excessive rain to the coastal plains of the southeastern U.S. These areas are also subject to heat and drought. Finding pasture crops that are able to withstand a variety of climatic assault proves challenging. When traditional forages aren't sustainable, less common forages become appealing. Bahiagrass (**BAH**), a warm-season perennial (*Paspalum notatum;* seeded spring, 2019), and reed canarygrass (**RCG**), a cool-season perennial (*Phalaris arundinacea;* seeded fall, 2019), were selected for renovating a pasture flooded twice in 5 years based on reports of flood tolerance in the literature. Seeded bermudagrass (**BER**, *Cynodon dactylon;* sown spring, 2019) was our control because of its success at other locations on the same farm. Stand vigor was checked periodically using the step-point technique. Prior to mowing in early June, 2020, samples were cut to estimate dry matter (**DM**) yield and nutrient content. Hay was harvested in July and September, 2020. Bale weights were recorded and bales cored for nutrient analysis. Plans are to overseed the grasses with clover in fall, 2020 to improve nutrient content of the overall swards and to continue monitoring productivity and sustainability. Step-point showed target grass represented an average of 47, 52, and 58% of the sward for RCG, BER, and BAH; respectively, over the growing season. Dry matter yield totaled 5,391; 7,781; and 9,923 lb/acre for RCG, BER, and BAH, respectively. Crude protein and total digestible nutrients (**TDN**) declined over the growing season. Crude protein and TDN averaged 7.9, 12.2, and 8.1% and 65.9, 68.2, and 59.1% of DM for BAH, RCG, and BER; respectively.

[1]Research/Extension Associate, (sfreeman@ncsu.edu); Beef Extension Specialist, (mhpoore@ncsu.edu), NC State Univ. Dept. of Animal Sci., Raleigh, NC.

WHITE LUPIN AND RYE AS POTENTIAL WINTER CROP FORAGES FOR THE SOUTHEAST

W.F. Anderson, L.L. Baxter, and E. Van Santen[1]

There is a constant search for high yielding, high quality forage production during the early spring months in the Southeast. White lupin (*Lupin alba* L.) is a legume that can supplement annual cereal rye (*Secale cereale* L.) or other spring crops for grazing, silage or haylage. White lupin is cultivated in many parts of the world as an edible food and as a high-protein forage for ruminants, especially cattle. White lupin as well as other lupins such as narrow-leaf *L. angustifolius* have been used as a winter cover to improve soil health. White lupin by itself and in combination with rye were grown at three location in Georgia (Tifton, Shellman, and Fort Valley) as a winter cover crop for the past three years. The covers were planted in late November after cotton and peanuts were harvested and residual mowed. Lupin and rye were planted with a strip-till rig and no-till planter in randomized complete block design with four replications. Narrow-leaf lupin with and without rye and rye alone were also included as treatments. Half of each of the plots were harvested and weighed. The other half of the plots were rolled and sprayed with glyphosate for preparation for planting of the summer row crops again. In the winter 2018-19 dry matter yields ranged from 2600 lb/acre in Tifton and Fort Valley Georgia to 7000 lb./acre in Shellman for white lupin alone. White lupin-rye mixtures had DM yields of 3300, 4130, and 8710 lb/acre at Tifton, Fort Valley, and Shellman respectively.

[1]Research Geneticist, USDA-ARS, Crop Genetics and Breeding Research Unit, Tifton, GA (bill.anderson@usda.gov); Forage Extension Specialist, University of Georgia, Tifton, GA (lbaxter@uga.edu); Statistical Consulting Unit, University of Florida, IFAS, Gainesville, GA (evsanten@ufl.edu).

YIELD GAINS BY INCORPORATING RHIZOMA PEANUT INTO BAHIAGRASS AND ITS IMPACT ON SOIL FERTILITY DISTRIBUTION

J.C Liu, C.L. Mackowiak, J.C.B. Dubeux Jr., H.L. Liao, and A.R. Blount[1]

Perennial peanut (*Arachis glabrata* Benth.) planted into bahiagrass (*Paspalum notatum* Flüggé) pastures provides N to the system, due to a symbiotic associating between RP and biological N_2 fixing (BNF) bacteria. Past studies focused on yield, forage quality and BNF. However, little is known on the impact of mixed plantings on forage nutrient composition or soil fertility. The following treatments were established in 2011: 1) monoculture 'Ecoturf' RP, 2) experimental Q6B RP, 3) monoculture 'Argentine' bahiagrass, 4) mixture of Ecoturf + bahiagrass, and 5) Q6B + bahiagrass. Forage and soil were sampled in July, 2020 and the forage measured for plant essential nutrients. Mehlich-3 extractable soil fertility was determined at three depth increments (0-15, 15-30, and 30-60 cm). Bahiagrass/RP mixtures yielded over 30% more forage than the bahiagrass or RP monocultures. Although soil fertility tended to decrease with soil depth, values were similar across treatments. However, legume tissue had increased P and K concentrations (%) when grown in mixtures with bahiagrass, while bahiagrass increased its Mg concentration in mixtures with RP. In contrast, the legumes tended to have lower Ca concentrations when grown as mixtures. Swards containing grass removed about 2x more P (kg/ha), while mixed swards removed about 1.5x more K. Swards with legumes removed 2x more Mg and over 4 times more Ca. Responses will continue to be tracked over time.

[1]Grad Student, Soil and Water Sciences, University of Florida (UF), Quincy, FL (jungchenliu@ufl.edu); Soil Specialist, Soil and Water Sciences, UF, Quincy, FL (echo13@ufl.edu); Forage Management Specialist, Agronomy, UF, Marianna, FL (dubeux@ufl.edu); Soil Microbiologist, Soil and Water Sciences, UF, Quincy, FL (sunny.liao@ufl.edu); Forage Breeder, Agronomy, UF, Quincy (paspalum@ufl.edu).

AUTHOR INDEX

American Forage and Grassland Council (AFGC)
PO Box 867
Berea, KY 40403
USA

ISBN 978-1-7138-2819-8

CURRAN ASSOCIATES INC.
proceedings.com